This is your very own math book! You can write in it, draw, circle, and color as you explore the exciting world of math.

Let's get started. Grab a crayon and draw a picture that shows what math means to you.

Have fun!

This is your space to draw.

Bothell, WA • Chicago, IL • Columbus, OH • New York, NY

connectED.mcgraw-hill.com

STEM McGraw-Hill is committed to providing instructional materials in Science, Technology, Engineering, and Mathematics (STEM) that give all students a solid foundation, one that prepares them for college and careers in the 21st century.

Send all inquiries to:
McGraw-Hill Education
STEM Learning Solutions Center
8787 Orion Place
Columbus, OH 43240

ISBN: 978-0-02-115021-2 *(Volume 1)*
MHID: 0-02-115021-4

Printed in the United States of America.

23 LWI 18

Our mission is to provide educational resources that enable students to become the problem solvers of the 21st century and inspire them to explore careers within Science, Technology, Engineering, and Mathematics (STEM) related fields.

Meet The Artists!

Alyssa Gonzalez

King of the Math Jungle This was a great experience similar to being on a roller coaster. Many friends and family supported me in this competition and many kids at Veterans Elementary will benefit from this. *Volume 1*

Finley Moss

Math is Yummy I came up with the idea because my mom and I love to bake cakes and it takes both time and measurement to do so. I was dreaming about how it would feel to win, so I am very excited! Math is yummy after all! *Volume 2*

Other Finalists

Landre Kate Beeler
Pop Numbers 4 Me

Sherry Bergeron's Class*
Math Makes Our World a Better Place

Sally Barmakian's Class*
Math is a Rainbow of Learning

Kamya Cooperwood
Monkey

Judson Upchurch
I like Numbers

Charles "Greyson" Biggs
Rainbow of Possibilities

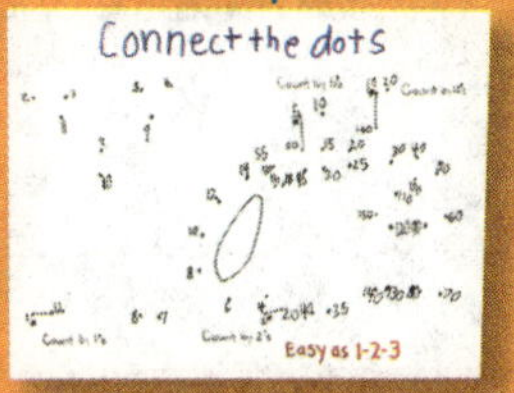

Leah Rauch
Connect the Dots

Andrew Morris
Math Designed World

Matthew Saldivat
Math Guy in the Sky

Ngun Za Cin
Three-Dimensional Shape Sculpture

Find out more about the winners and other finalists at www.MHEonline.com.

We wish to congratulate all of the entries in the 2011 *McGraw-Hill My Math* "What Math Means To Me" cover art contest. With over 2,400 entries and more than 20,000 community votes cast, the names mentioned above represent the two winners and ten finalists for this grade.

** Please visit mhmymath.com for a complete list of students who contributed to this artwork.*

it's all at
connectED.mcgraw-hill.com

Go to the Student Center for your eBook, Resources, Homework, and Messages.

Write your Username

Password

Get your resources online to help you in class and at home.

Vocab

Find activities for building vocabulary.

Watch

Watch animations of key concepts.

Tools

Explore concepts with virtual manipulatives.

Check

Self-assess your progress.

eHelp

Get targeted homework help.

Games

Reinforce with games and apps.

Tutor

See a teacher illustrate examples and problems.

GO mobile

Scan this QR code with your smart phone* or visit mheonline.com/stem_apps.

*May require quick response code reader app.

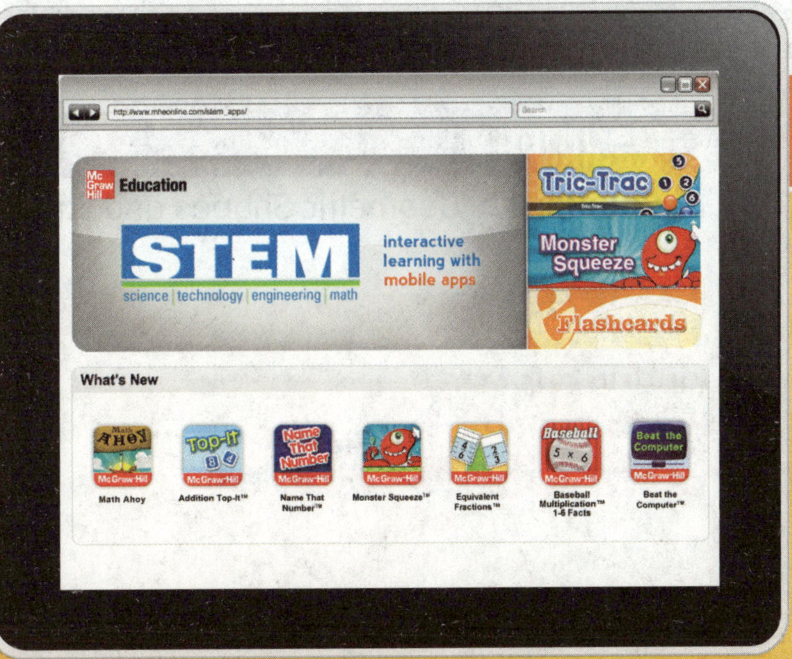

Contents in Brief

Organized by Domain

Chapter 1

Apply Addition and Subtraction Concepts

ESSENTIAL QUESTION
What strategies can I use to add and subtract?

Getting Started

Lessons and Homework

Wrap Up

connectED.mcgraw-hill.com

Look for this!
Watch
Click online and you can watch videos that will help you learn the lessons.

Chapter 2 Number Patterns

ESSENTIAL QUESTION
How can equal groups help me add?

Getting Started

Lessons and Homework

Wrap Up

Hi!

connectED.mcgraw-hill.com

Chapter 3 Add Two-Digit Numbers

ESSENTIAL QUESTION
How can I add two-digit numbers?

Getting Started

Lessons and Homework

Wrap Up

We make a great pair!

connectED.mcgraw-hill.com

Look for this!
Click online and you can get more help while doing your homework.

Chapter 4

Subtract Two-Digit Numbers

ESSENTIAL QUESTION
How can I subtract two-digit numbers?

Getting Started

Lessons and Homework

Wrap Up

connectED.mcgraw-hill.com

Chapter 5 Place Value to 1,000

ESSENTIAL QUESTION
How can I use place value?

Getting Started

Lessons and Homework

Wrap Up

connectED.mcgraw-hill.com

Look for this! Click online and you can find tools that will help you explore concepts.

Chapter 6 Add Three-Digit Numbers

ESSENTIAL QUESTION
How can I add three-digit numbers?

Getting Started

Lessons and Homework

Wrap Up

connectED.mcgraw-hill.com

Chapter 7 Subtract Three-Digit Numbers

ESSENTIAL QUESTION
How can I subtract three-digit numbers?

Getting Started

Lessons and Homework

Wrap Up

Look for this!
Click online and you can find activities to help build your vocabulary.

You + School = Cool

connectED.mcgraw-hill.com

Chapter 8
Money

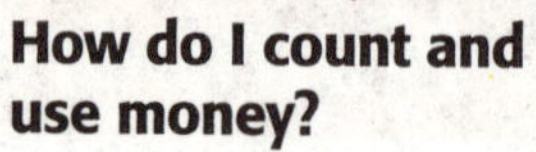

ESSENTIAL QUESTION
How do I count and use money?

Getting Started

Lessons and Homework

Wrap Up

Chapter 9 Data Analysis

ESSENTIAL QUESTION
How can I record and analyze data?

Getting Started

Go nuts!

Lessons and Homework

Wrap Up

connectED.mcgraw-hill.com

Chapter 10 Time

ESSENTIAL QUESTION
How do I use and tell time?

Getting Started

Lessons and Homework

Wrap Up

connectED.mcgraw-hill.com

Chapter 11 Customary and Metric Lengths

ESSENTIAL QUESTION
How can I measure objects?

I love rulers because I'm an inch worm!

Getting Started

Lessons and Homework

Wrap Up

connectED.mcgraw-hill.com

Chapter 12 Geometric Shapes and Equal Shares

ESSENTIAL QUESTION
How do I use shapes and equal parts?

Getting Started

Lessons and Homework

Wrap Up

connectED.mcgraw-hill.com

Chapter 1

Apply Addition and Subtraction Concepts

ESSENTIAL QUESTION

What strategies can I use to add and subtract?

Let's Find Some Animals!

Watch a video!

Watch

My Common Core State Standards

Operations and Algebraic Thinking

2.OA.1 Use addition and subtraction within 100 to solve one- and two-step word problems involving situations of adding to, taking from, putting together, taking apart, and comparing, with unknowns in all positions.

2.OA.2 Fluently add and subtract within 20 using mental strategies. By end of Grade 2, know from memory all sums of two one-digit numbers.

Number and Operations in Base Ten *This chapter also addresses these standards:*

2.NBT.5 Fluently add and subtract within 100 using strategies based on place value, properties of operations, and/or the relationship between addition and subtraction.

2.NBT.9 Explain why addition and subtraction strategies work, using place value and the properties of operations.

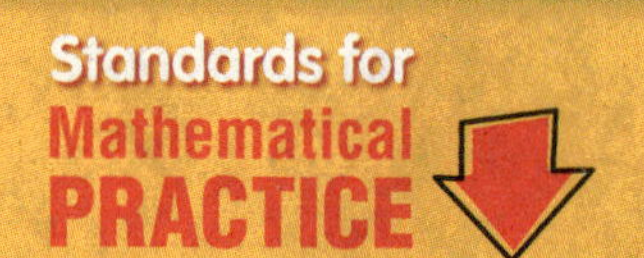

1. Make sense of problems and persevere in solving them.
2. Reason abstractly and quantitatively.
3. Construct viable arguments and critique the reasoning of others.
4. Model with mathematics.
5. Use appropriate tools strategically.
6. Attend to precision.
7. Look for and make use of structure.
8. Look for and express regularity in repeated reasoning.

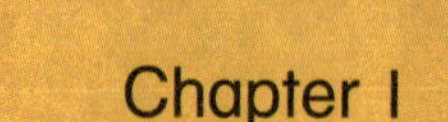
= focused on in this chapter

Name ....

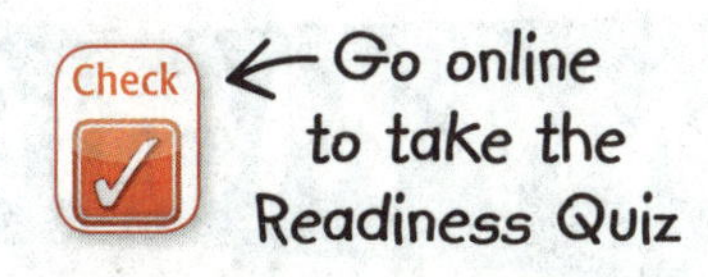

Add.

1. + ______ bananas

Subtract.

2. $9 - 7 =$ ______

Look at the picture. Write the number sentence. Add.

3.

______ ◯ ______ ◯ ______ apples

4. Christy put 6 marbles in a group. 2 marbles rolled away. How many marbles are left?

______ marbles

How Did I Do? **Shade the boxes to show the problems you answered correctly.**

1	2	3	4

Name

My Math Words

Review Vocabulary

minus sign (–)   plus sign (+)

Write the word *addition* or *subtraction* in the center. Then use a review vocabulary word to describe your word.

Show as a Symbol

Show as a Number Sentence

Show an Example Using a Drawing

Show a Non-Example Using a Drawing

My Vocabulary Cards

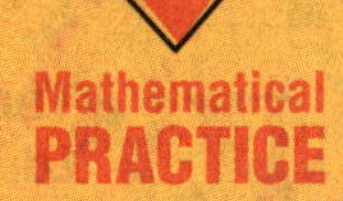

Lesson 1-1

add

4 + 3 = 7

Lesson 1-1

addend

6 + 4 = 10

Lesson 1-7

count back

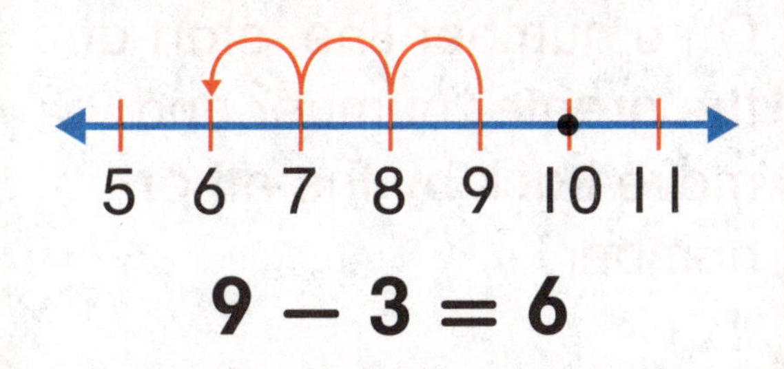

9 – 3 = 6

Lesson 1-2

count on

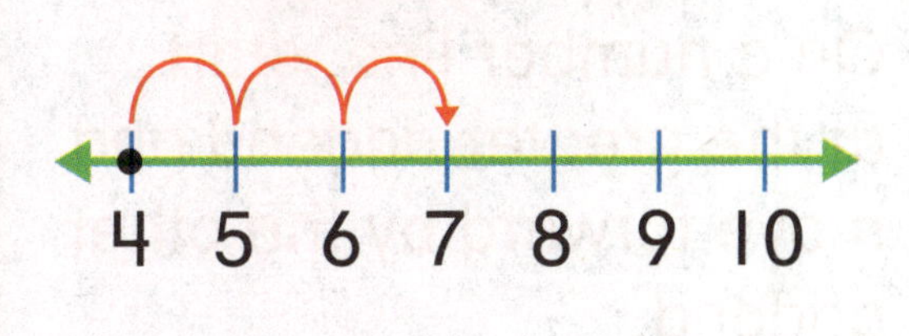

4 + 3 = 7

Lesson 1-7

difference

5 – 2 = 3

Lesson 1-3

doubles

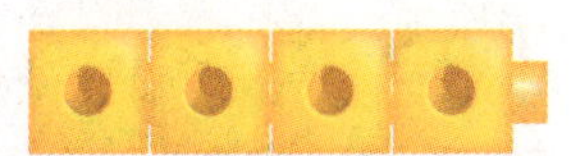

4 + 4 = 8

Teacher Directions: Ideas for Use

- Ask students to arrange cards to show an opposite pair. Have them explain the meaning of their paring.
- Write a tally mark on each card every time you read the word in this chapter or use it in your writing. Try to use at least 10 tally marks for each word card.

Any numbers being added together.

To join together sets to find the total or sum. The opposite of *subtract*.

On a number line, start at the greater addend and move forward by the other addend.

On a number line, start at the greater number and move back by the other number.

Two addends that are the same.

The answer to a subtraction problem.

My Vocabulary Cards

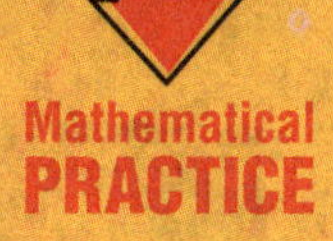

Mathematical PRACTICE

Lesson 1-12

fact family

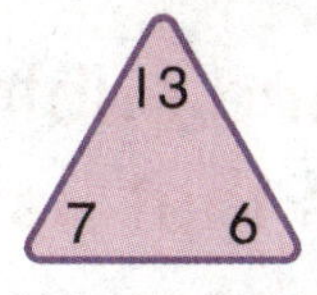

$6 + 7 = 13$ $13 - 6 = 7$
$7 + 6 = 13$ $13 - 7 = 6$

Lesson 1-11

missing addend

$5 + \square = 9$
$9 - 5 = \square$

Lesson 1-3

near doubles

$5 + 6 = 11$

Lesson 1-10

related facts

$3 + 4 = 7$ $4 + 3 = 7$
$7 - 4 = 3$ $7 - 3 = 4$

Lesson 1-7

subtract

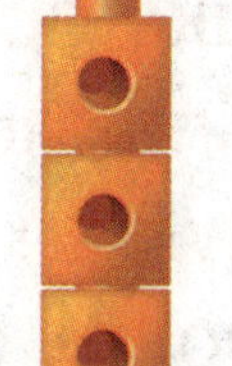

$5 - 2 = 3$

Lesson 1-1

sum

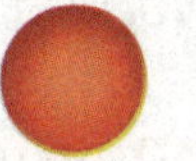

$2 + 4 = 6$

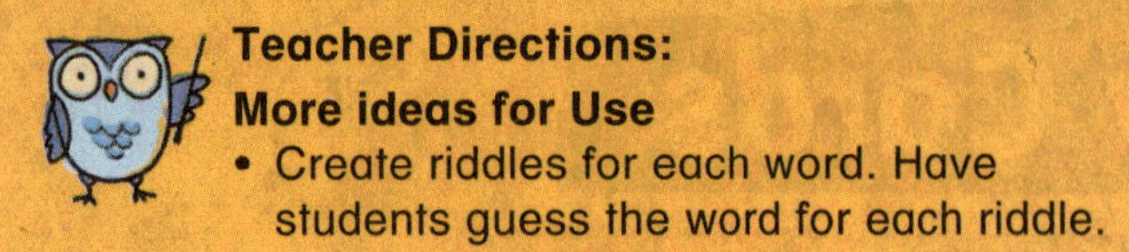

Teacher Directions:

More ideas for Use

- Create riddles for each word. Have students guess the word for each riddle.
- Have students draw examples for each card. Have them make drawings that are different from what is shown on each card.

The missing number in a number sentence that makes the number sentence true.

Addition and subtraction sentences that use the same numbers.

Basic facts using the same number. Sometimes called a *fact family*.

Addition facts in which one addend is exactly 1 more or 1 less than the other addend.

The answer to an addition problem.

Take away, take apart, separate, or find the difference between two sets. The opposite of *add*.

My Foldable

FOLDABLES® Follow the steps on the back to make your Foldable.

2 + 1

6 + 4

8 + 7

3 + 9

7 + 5

4 + 0

___ ◯ ___

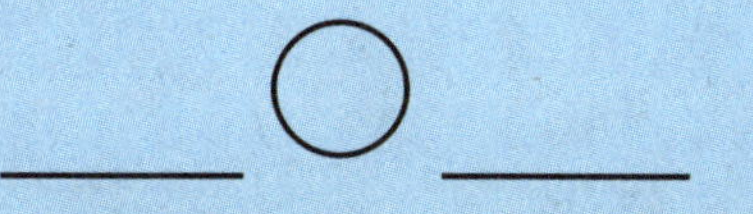

1

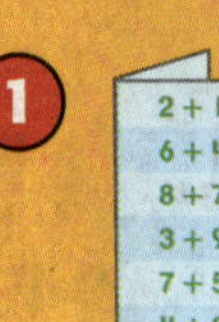

2

2 + 1
6 + 4
8 + 7
3 + 9
7 + 5
4 + 0

3

$1 + 2 =$ 3

___ ○ ___ = ___

___ ○ ___ = ___

___ ○ ___ = ___

___ ○ ___ = ___

___ ○ ___ = ___

___ ○ ___ = ___

___ ○ ___ = ___

Name ....

Addition Properties

Lesson 1

ESSENTIAL QUESTION
What strategies can I use to add and subtract?

Explore and Explain

6 + 2 = 8

9 + 5 = 14

Teacher Directions: Draw a picture and write a number sentence to solve. Sue and Pat are zookeepers. Sue fed 5 monkeys and 3 elephants. How many animals did she feed? Pat fed 3 tigers and 5 giraffes. How many animals did he feed?

See and Show

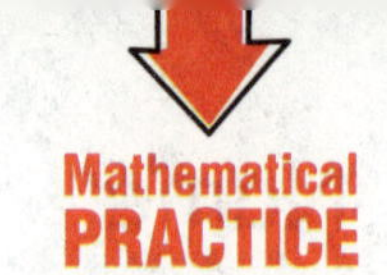

You **add** to find a sum. Each number you add is called an **addend**. The answer is the **sum**.

Helpful Hint
You can add numbers in any order. The sum is the same. This is the Commutative Property.

3 + 5 = 8 (sum)
addends

5 + 3 = 8 (sum)
addends

Any number plus zero equals that number. This is the Identity Property.

5 + 0 = 5
(5 and 0: addends; 5: sum)

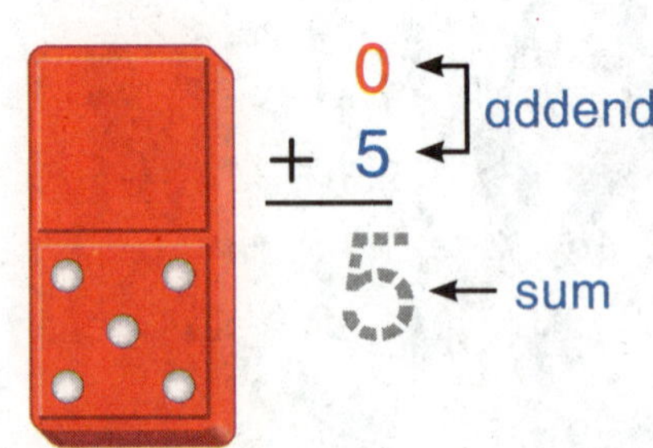

0 + 5 = 5
(0 and 5: addends; 5: sum)

Find each sum.

1. 4 + 3 = 7

3 + 4 = 7

2. 

6 + 0 = 6

0 + 6 = 6

3. 

2 + 4 = 6

4 + 2 = 6

4. 

2 + 5 = 7

5 + 2 = 7

Talk Math Why is the sum the same when you find 3 + 2 or 2 + 3?

Name ....

CCSS

On My Own

Find each sum.

5. $5 + 1 = \underline{6}$

$1 + 5 = \underline{6}$

6.

$\begin{array}{r} 4 \\ +\ 5 \\ \hline 9 \end{array}$ $\begin{array}{r} 5 \\ +\ 4 \\ \hline 9 \end{array}$

7. $\begin{array}{r} 0 \\ +\ 3 \\ \hline 3 \end{array}$ $\begin{array}{r} 3 \\ +\ 0 \\ \hline 3 \end{array}$

8. $\begin{array}{r} 6 \\ +\ 3 \\ \hline 9 \end{array}$ $\begin{array}{r} 3 \\ +\ 6 \\ \hline 9 \end{array}$

9. $\begin{array}{r} 7 \\ +\ 1 \\ \hline 8 \end{array}$ $\begin{array}{r} 1 \\ +\ 7 \\ \hline 8 \end{array}$

10. $6 + 2 = \underline{8}$

$\underline{8} = 2 + 6$

11. $8 + 0 = \underline{8}$

$0 + 8 = \underline{8}$

12. $\underline{6} = 4 + 2$

$2 + 4 = \underline{6}$

13. $3 + 4 = \underline{7}$

$4 + 3 = \underline{7}$

14. $0 + 9 = \underline{9}$

$9 + 0 = \underline{9}$

Problem Solving

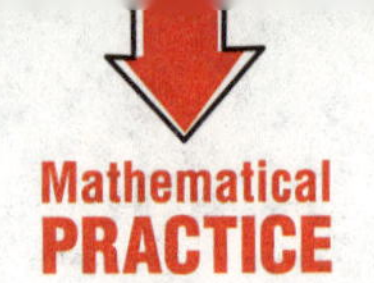

15. What two addition facts could you use to find the total number of dots on this domino?

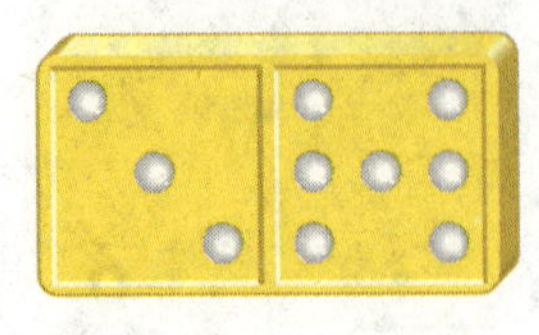

_____ + _____ = _____   _____ + _____ = _____

16. Manuel's team scores 𝍸| runs in the first game. They score ||| runs in the second game. Show two ways you can find the total number of runs.

_____ + _____ = _____   _____ + _____ = _____

Write Math Write what you know about the order of the addends in an addition problem.

____

____

____

Name Bri

Count On to Add

Lesson 2
ESSENTIAL QUESTION
What strategies can I use to add and subtract?

Explore and Explain

Teacher Directions: Draw a green box around number 1. Count on 3. Draw a purple box around that number. Draw a green box around number 7. Count on 2. Draw a purple box around that number. Draw a green box around number 11. Count on 1. Draw a purple box around that number.

See and Show

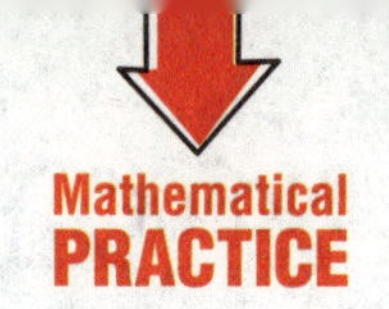

Find 7 + 3. Use a number line. **Count on** to add.
Start with the greater addend.

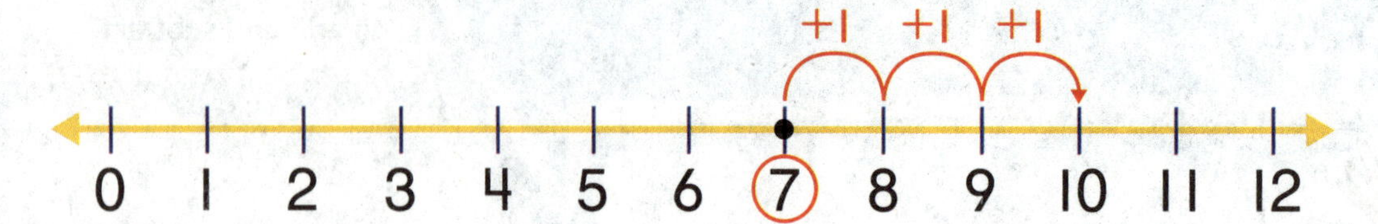

Start at 7. Count on 3.

7 + 3 = 10

10 = 7 + 3

Count on to add. Use the number line to help.

1. 6 + 3 = 9
2. 6 = 4 + 2
3. 2 + 9 = 11
4. 1 + 3 = 4
5. 9 = 2 + 7
6. 10 = 9 + 1

Talk Math Why should you count on from the greater addend?

CCSS

Name ....

On My Own

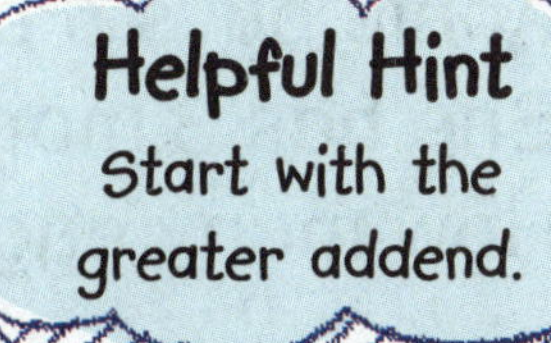

Count on to add. Use the number line to help.

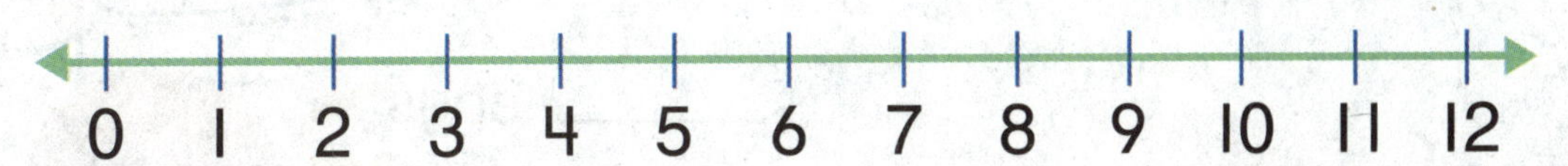

7. $7 + 1 =$ 8

8. 7 $= 5 + 2$

9. $3 + 8 =$ 11

10. $6 + 1 =$ 7

11. 12 $= 9 + 3$

12. $2 + 8 =$ 10

13. $2 + 7$ = 9

14. $3 + 2$ = 5

15. $1 + 9$ = 10

16. $5 + 1$ = 6

17. $3 + 6$ = 9

18. $1 + 4$ = 5

19. $3 + 1$ = 4

20. $2 + 6$ = 8

21. $8 + 1$ = 9

Problem Solving

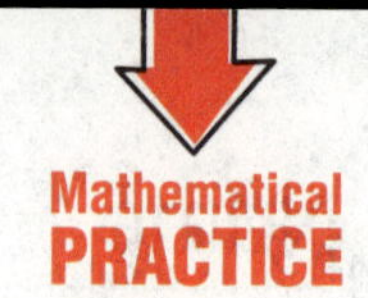

22. Annie buys 6 eggs at the market. She has 3 more eggs at home. How many eggs does she have in all?

9 eggs

23. Sal's cow gave 3 pails of milk in the morning and 5 pails in the afternoon. How many pails of milk did Sal's cow give?

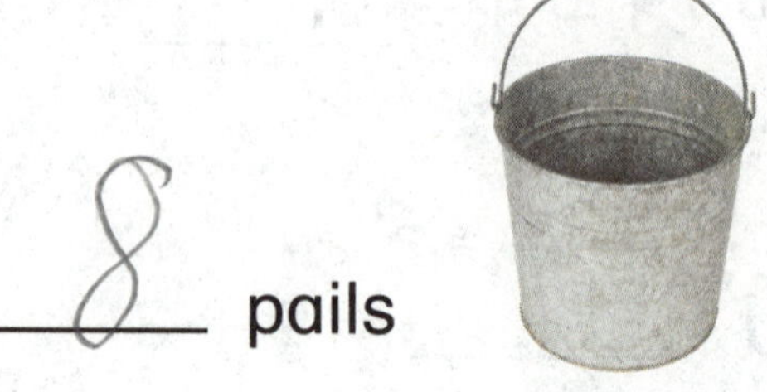

8 pails

24. Joseph has 4 pigs on his farm. One pig has 3 piglets. How many pigs are on the farm?

7 pigs

Write Math How does counting on help you add?

____

____

____

Name ____________________

My Homework

Lesson 2
Count On to Add

Homework Helper

Need help? connectED.mcgraw-hill.com

Use a number line to find 9 + 3. Start with the greater addend. Count on to add.

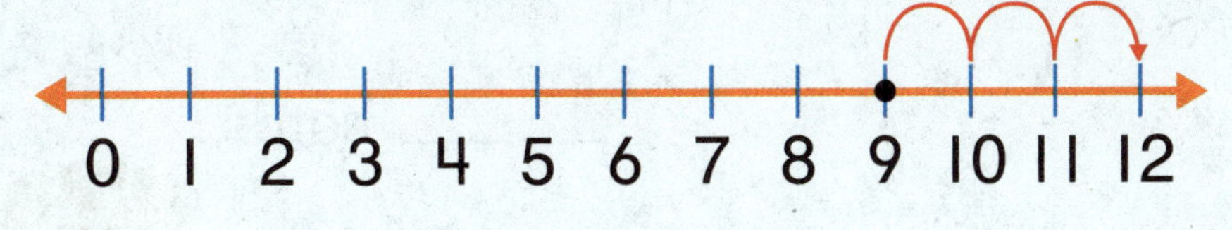

9 + 3 = 12

Practice

Count on to add. Use the number line above.

1. 4 + 1 = ______

2. 2 + 6 = ______

3. 1 + 8 = ______

4. 3 + 6 = ______

5. 2 + 4 = ______

6. 2 + 8 = ______

7. $\begin{array}{r} 3 \\ +\ 9 \\ \hline \end{array}$

8. $\begin{array}{r} 4 \\ +\ 3 \\ \hline \end{array}$

9. $\begin{array}{r} 1 \\ +\ 4 \\ \hline \end{array}$

Read and then solve the problems.

10. 6 bees are buzzing near a hive.
3 more bees come out of the hive.
How many bees are there now?

______ bees

11. Cherie has 5 trading cards. She gets 3 more cards. How many cards does Cherie have now?

______ cards

I love baseball!

12. There are 8 butterflies on a bush. 3 more butterflies fly onto the bush. How many butterflies are on the bush now?

______ butterflies

Vocabulary Check

Draw lines to match.

13. count on

$6 + 3 = 9$

14. addend

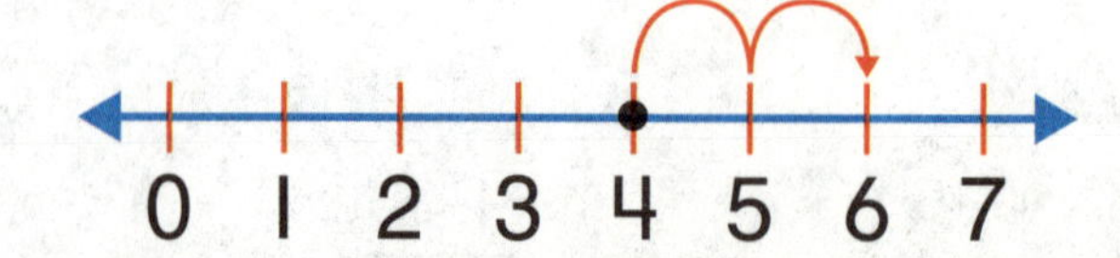

Math at Home Give your child an addition fact. Have him or her count on from the greater number to find the sum.

Name ....

Doubles and Near Doubles

Lesson 3

ESSENTIAL QUESTION

What strategies can I use to add and subtract?

I am seeing doubles!

Explore and Explain

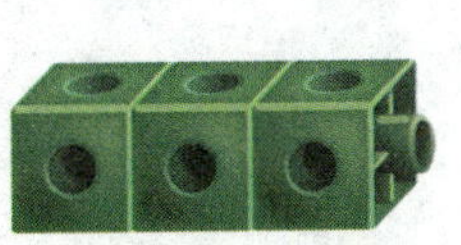

_____ + _____ = _____

_____ + _____ = _____

_____ + _____ = _____

Teacher Directions: Look at each cube train. Draw the same train beside it. Write the doubles fact shown by each pair of cube trains.

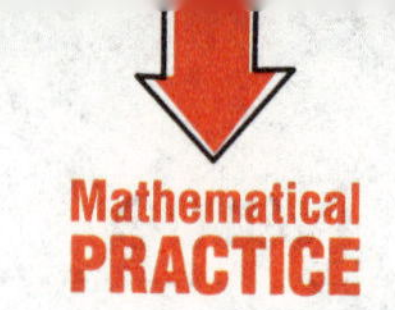

See and Show

Two addends that are the same are called **doubles**.

Use doubles facts to find the sum.

6 + 6 = 12

addend, addend, sum

Two addends that are almost a doubles fact are called **near doubles**.

Use near doubles facts to find the sum.

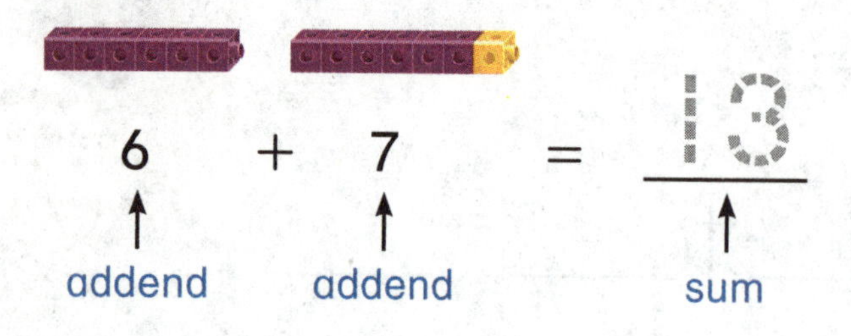

6 + 7 = 13

addend, addend, sum

Add. Circle the doubles facts.

1. 4 + 4 = ______

2.  3 + 4 = ______

3. ______ = 7 + 7

4. ______ = 9 + 9

5. 7 + 8 = ______

6. ______ = 8 + 8

Talk Math How can you use doubles and near doubles to remember 5 + 5? 5 + 6?

Name ....

On My Own

Add. Circle the doubles facts.

7. 0 + 0 = ______

8. 7 + 6 = ______

9. 2 + 8 = ______

10. 8 + 9 = ______

11. 3 + 2 = ______

12. 2 + 9 = ______

13. 2 + 2 = ______

14. 7 + 7 = ______

15. 1 + 1 = ______

16. 9 + 3 = ______

Add. Circle the near doubles facts.

17. 4 + 3 = ______

18. 2 + 4 = ______

19. 7 + 8 = ______

20. 3 + 8 = ______

21. 2 + 7 = ______

22. 9 + 8 = ______

23. 3 + 5 = ______

24. 4 + 5 = ______

Problem Solving

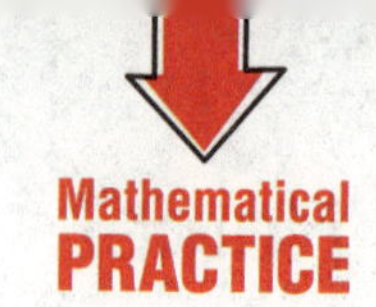

25. Mr. Bean sells 5 melons to Ed. He sells the same number of melons to Jose. How many melons did Mr. Bean sell in all?

_____ melons

26. Andrea has 3 dogs. Juanita has the same number of dogs. How many dogs do they have in all?

_____ dogs

27. Anthony has 3 lizards. His cousin has one more lizard than he does. How many lizards do Anthony and his cousin have in all?

_____ lizards

How does knowing doubles facts help you with solving near doubles facts?

____

____

____

Name

My Homework

Lesson 3
Doubles and Near Doubles

Homework Helper

Need help? 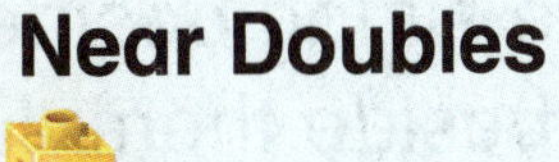

Two addends that are the same are called doubles.

Two addends that are almost a doubles fact are called near doubles.

Doubles

4 + 4 = 8

Near Doubles

4 + 5 = 9

Practice

Add. Circle the doubles facts.

1. 5 + 7 = ______

2. 2 + 3 = ______

3. 5 + 5 = ______

4. 8 + 7 = ______

5. 4 + 2

6. 4 + 4

7. 6 + 7

8. 8 + 8

Add. Circle the near doubles facts.

9. $2 + 9 =$ ______

10. $12 + 3 =$ ______

11. $4 + 5 =$ ______

12. $8 + 8 =$ ______

13. $9 + 8 =$ ______

14. $7 + 6 =$ ______

Solve. Then write the doubles fact that helped you solve the problem.

15. 4 frogs are sitting beside a pond. 5 dragon flies land beside them. How many animals are there altogether?

_____ + _____ = _____ animals

_____ + _____ = _____

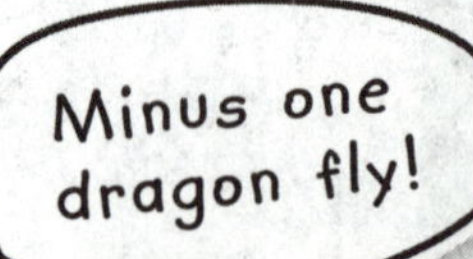

Vocabulary Check

Draw lines to match.

16. **near doubles** $4 + 4 = 8$

17. **doubles** $4 + 5 = 9$

Math at Home Have your child use objects to make doubles and tell the addition fact.

Name ______________________

Make a 10

Lesson 4

ESSENTIAL QUESTION
What strategies can I use to add and subtract?

Explore and Explain

You can count on us!

____ + ____ = ____

____ + ____ = ____

Teacher Directions: Place 13 counters in a cup. Empty the cup onto the page. Place red counters in the top ten-frame. Place yellow counters in the bottom ten-frame. Write the addition sentence. Move yellow counters into the top ten-frame to make ten. Write the new addition sentence.

See and Show

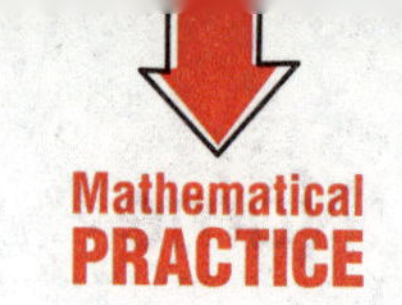

You can make a ten to help you add. Find 8 + 4.

First:
Show 8.
Then show 4.

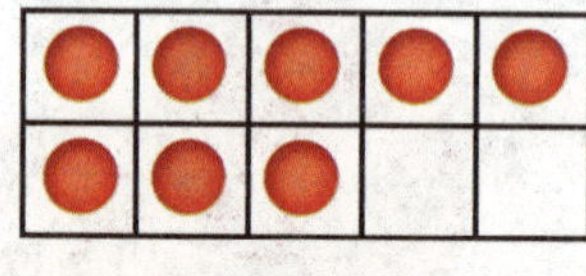

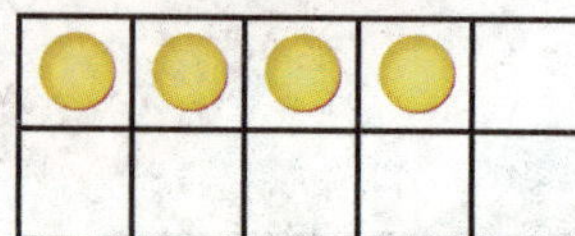

8 + 4

Next:
Take apart 4
to make a 10.

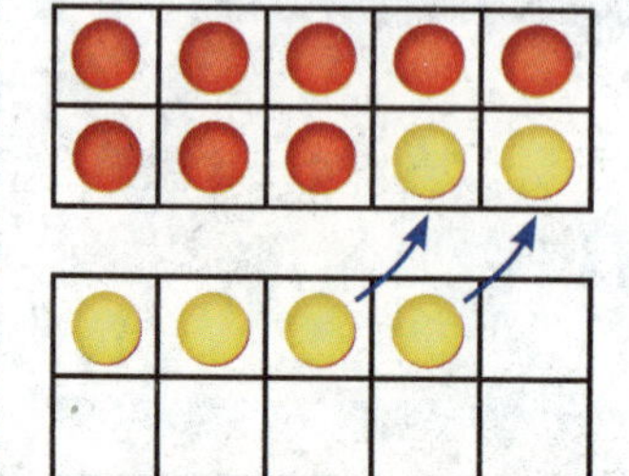

8 + 4
2 2

8 + 2 = 10

Last:
Add.

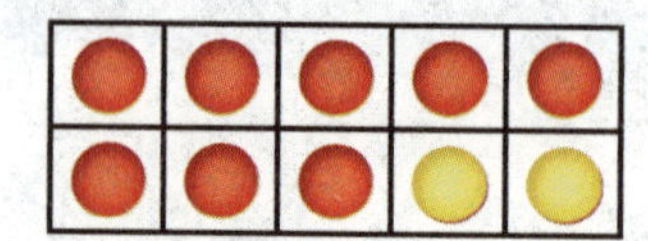

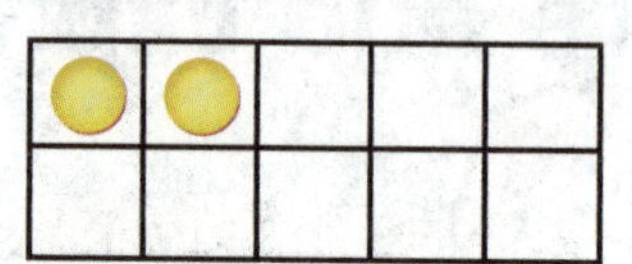

Now you have 10 + 2.

10 + 2 = 12

So, 8 + 4 = 12.

Use Work Mat 2 and . Make a 10 to add.

1.

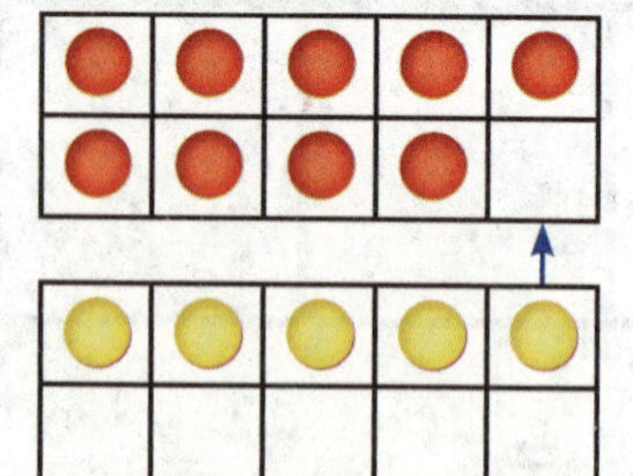

9 + 5 = 14

2.

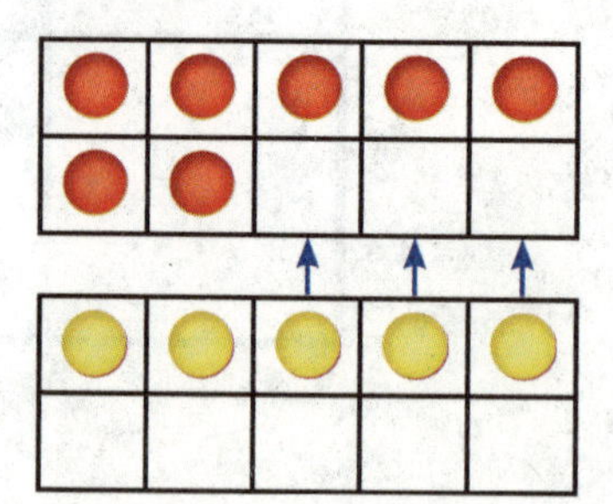

7 + 5 = 12

Talk Math Name all of the facts you know that have a sum of 10.

Name

CCSS

On My Own

Use Work Mat 2 and ●●. Make a 10 to add.

3. 8 + 6 = 14

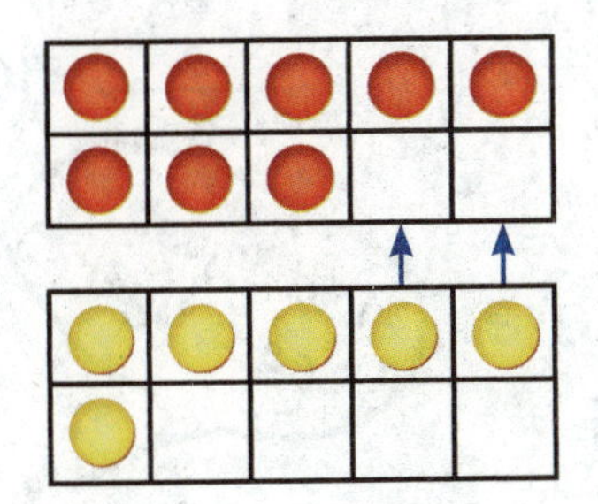

4. 7 + 8 = 15

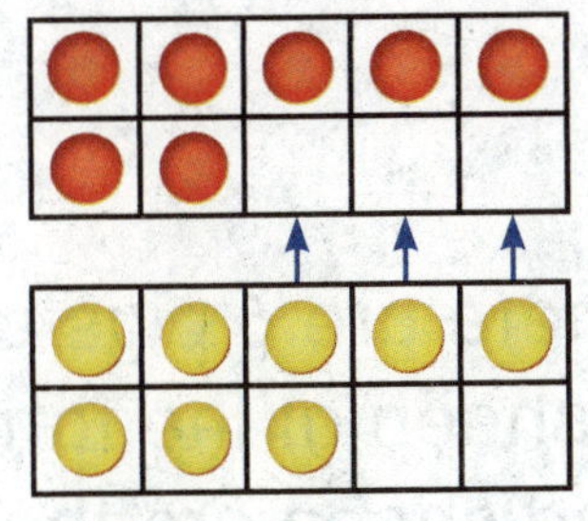

5. 9 + 4 = 13

6. 7 + 7 = 14

7. 4 + 8 = 12

8. 8 + 9 = 17

9. 3 + 9 = 12

10. 7 + 5 = 12

11. $\begin{array}{r} 3 \\ +\ 7 \\ \hline 10 \end{array}$

12. $\begin{array}{r} 4 \\ +\ 9 \\ \hline 13 \end{array}$

13. $\begin{array}{r} 2 \\ +\ 9 \\ \hline 11 \end{array}$

14. $\begin{array}{r} 7 \\ +\ 8 \\ \hline 15 \end{array}$

15. $\begin{array}{r} 8 \\ +\ 4 \\ \hline \end{array}$

16. $\begin{array}{r} 4 \\ +\ 6 \\ \hline \end{array}$

Problem Solving

Mathematical PRACTICE

17. Ms. Ling's class sees 9 tigers at the zoo. Ms. John's class sees 8 tigers. How many tigers did they see in all?

_____ tigers

18. There are 6 pigs at the farm. There are 7 sheep at the farm. How many pigs and sheep are there in all?

_____ pigs and sheep

19. 5 goldfish are in the tank. 7 more goldfish are added to the tank. How many goldfish are in the tank now?

_____ goldfish

How could you use making a ten when adding?

____

____

____

Name ....

My Homework

Lesson 4
Make a 10

Homework Helper

Need help? connectED.mcgraw-hill.com

Make a ten to help you add.

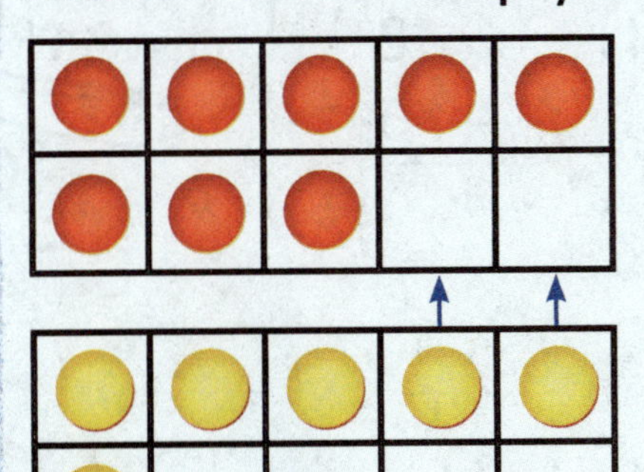

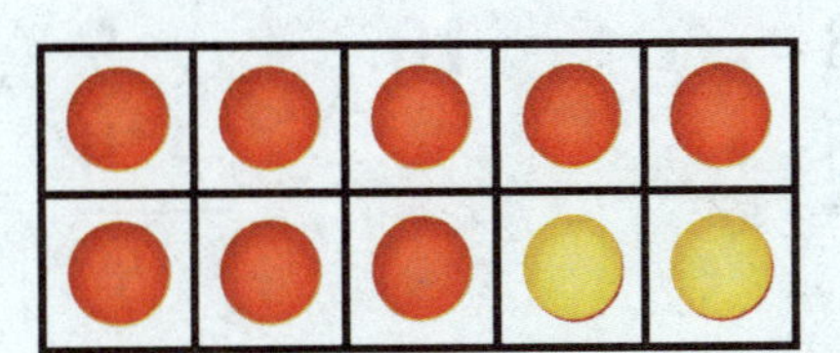

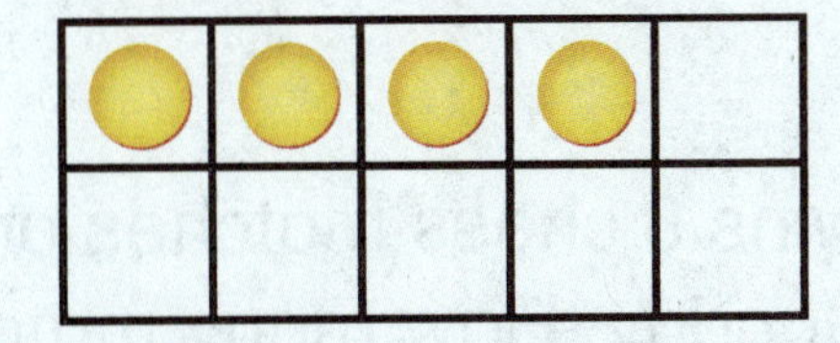

8 + 6

Think: 8 + 2 = 10
Add: 10 + 4 = 14.
So, 8 + 6 = 14.

Practice

Make a ten to add.

1.

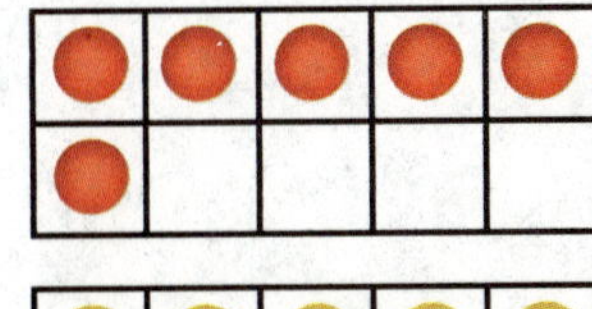

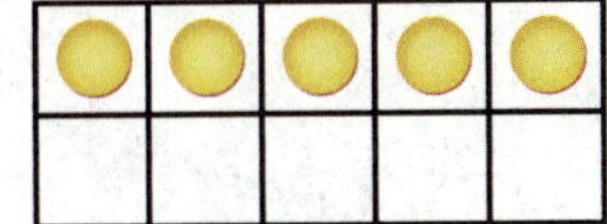

6 + 5 = ______

2.

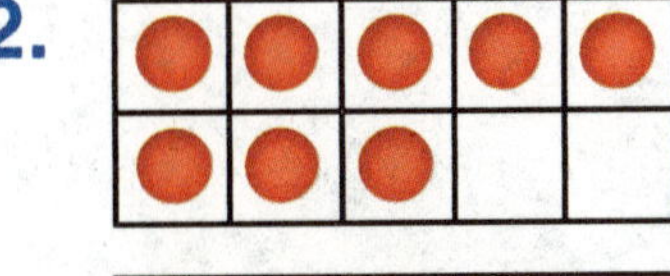

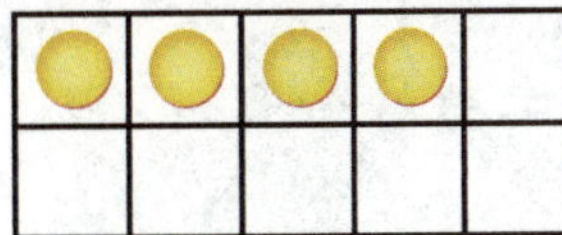

8 + 4 = ______

Make a ten to add.

3. $7 + 4 =$ _______

4. $4 + 8 =$ _______

5. $9 + 7 =$ _______

6. $7 + 6 =$ _______

7. $3 + 9 =$ _______

8. $7 + 5 =$ _______

9. $\begin{array}{r} 8 \\ +\ 8 \\ \hline \end{array}$

10. $\begin{array}{r} 9 \\ +\ 4 \\ \hline \end{array}$

11. $\begin{array}{r} 8 \\ +\ 5 \\ \hline \end{array}$

12. Raul wins 8 chess matches on Saturday and 5 on Sunday. How many matches did he win?

_____ matches

Test Practice

13. Which addition sentence can help you find the sum for $8 + 7$?

$5 + 7 = 12$ ○	$10 + 5 = 15$ ○
$10 + 8 = 18$ ○	$10 + 7 = 17$ ○

Math at Home Have your child tell you how to find $8 + 7$ by making a 10.

Name ....

Add Three Numbers

Lesson 5

ESSENTIAL QUESTION
What strategies can I use to add and subtract?

Explore and Explain

_____ animals

Teacher Directions: Have students use counters to show the following: There are 4 fish, 6 frogs, and 2 snakes in the pond. Trace the counters and write how many in all.

See and Show

Mathematical PRACTICE

Helpful Hint
Look for facts you know.

You can group addends in different ways. The sum will stay the same.

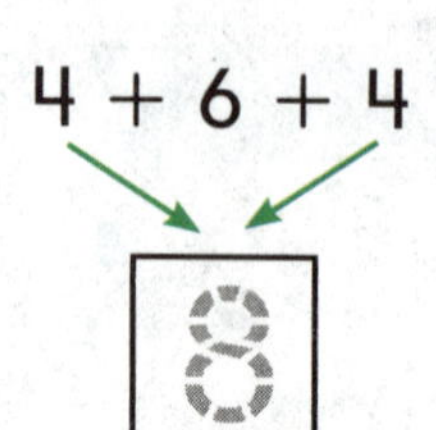

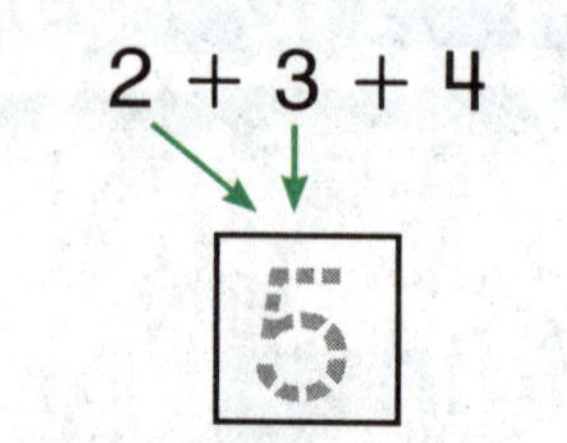

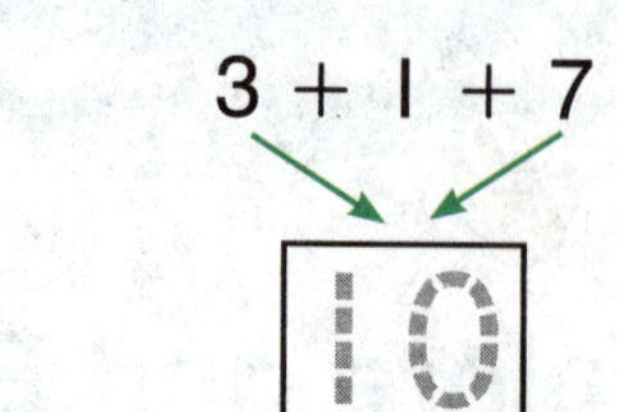

6 + 8 = 14   5 + 4 = 9   1 + 10 = 11

Find each sum. Circle the numbers you add first. Write that sum in the box.

1. 2 + 4 + 2 = ______

2. 9 + 1 + 2 = ______

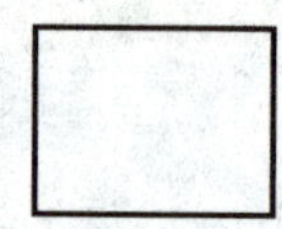

3. 7 + 4 + 7 = ______

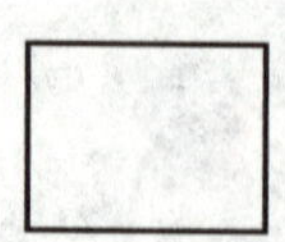

4. 2 + 7 + 8 = ______

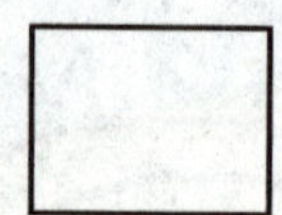

5. 3 + 7 + 4 = ______

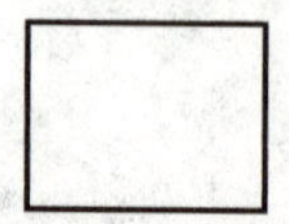

6. 5 + 3 + 5 = ______

Talk Math How do you decide which two numbers to add first when adding three digits?

Name ....

CCSS

Helpful Hint
Add two numbers first. Look for facts you know.

On My Own

Find each sum. Circle the numbers you add first. Write that sum in the box.

7. 4 + 3 + 4 = ______

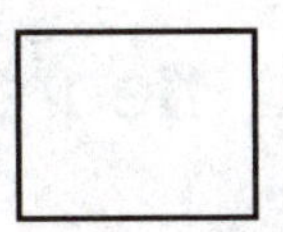

8. 4 + 3 + 7 = ______

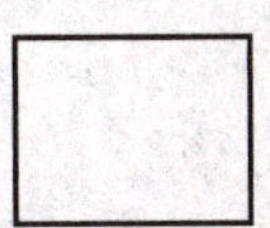

9. 2 + 8 + 3 = ______

10. 6 + 1 + 6 = ______

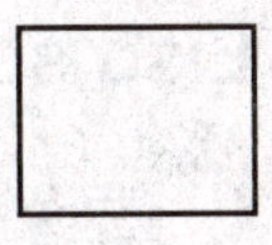

Find each sum.

11. 5 + 5 + 5 = ______

12. 6 + 6 + 3 = ______

13. 7 + 4 + 7 = ______

14. 9 + 8 + 1 = ______

15. 1 + 7 + 3 = ______

16. 5 + 7 + 5 = ______

17. $\begin{array}{r} 3 \\ 5 \\ +\ 7 \\ \hline \end{array}$

18. $\begin{array}{r} 7 \\ 7 \\ +\ 1 \\ \hline \end{array}$

19. $\begin{array}{r} 9 \\ 8 \\ +\ 1 \\ \hline \end{array}$

Problem Solving

Mathematical PRACTICE

20. The zoo has 5 black bears, 5 brown bears and 2 polar bears. How many bears are at the zoo?

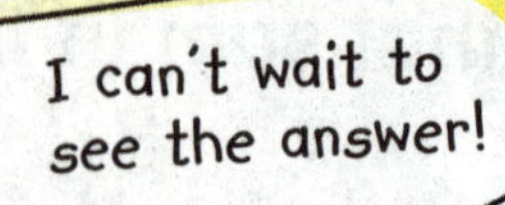

______ bears

21. Lisa saw 4 sheep and 4 goats at the petting zoo. She also saw 3 piglets. How many animals did Lisa see at the petting zoo?

______ animals

Write the doubles fact that helped you solve the problem.

______ + ______ = ______

Write Math Explain how you can group addends in any way to help you solve addition problems that have more than 2 addends.

____

____

____

Name ....

Operations and Algebraic Thinking
2.OA.1, 2.OA.2

CCSS

My Homework

Lesson 5
Add Three Numbers

Homework Helper

Need help? connectED.mcgraw-hill.com

Look for facts you know to help you add three numbers.

6 + 5 + 4 → 6 + 4 = 10 → 10 + 5 = 15

4 + 5 + 4 → 4 + 4 = 8 → 8 + 5 = 13

3 + 1 + 4 → 3 + 1 = 4 → 4 + 4 = 8

Practice

Find each sum.

1. 6 + 6 + 4 = ______

2. 6 + 2 + 8 = ______

3. 3 + 3 + 9 = ______

4. 7 + 4 + 3 = ______

5. 1 + 9 + 4 = ______

6. 7 + 6 + 6 = ______

7. 8 + 4 + 2 = ______

Find each sum.

8.
$$\begin{array}{r} 7 \\ 3 \\ +\ 5 \\ \hline \end{array}$$

9.
$$\begin{array}{r} 4 \\ 2 \\ +\ 6 \\ \hline \end{array}$$

10.
$$\begin{array}{r} 9 \\ 8 \\ +\ 1 \\ \hline \end{array}$$

11. The zoo has 3 giraffes, 4 African elephants, and 3 Asian elephants. How many elephants and giraffes are at the zoo?

_____ giraffes and elephants

12. There are 5 tigers, 4 leopards, and 6 lions at the zoo. How many big cats are at the zoo?

_____ big cats

Test Practice

13. Which addition sentence could help solve this problem?

$3 + 5 + 7 =$ _____

○ $2 + 3 = 5$ ○ $5 + 5 = 10$ ○ $3 + 7 = 10$ ○ $5 + 2 = 7$

Math at Home Have your child show you how to add 7 + 7 + 1.

Name

Problem Solving

STRATEGY: Write a Number Sentence

Lesson 6

ESSENTIAL QUESTION
What strategies can I use to add and subtract?

Watch

Four friends were eating a snack. Two more friends joined them. Each person ate a banana. How many bananas did they eat?

Don't slip!

1 Understand

Underline what you know.
Circle what you need to find.

2 Plan

How will I solve the problem?

3 Solve

Write a number sentence.

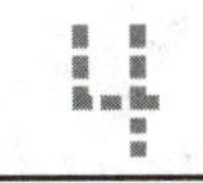   

4 + 2 = 6

6 bananas

Too late!

4 Check

Is my answer reasonable? Explain.

Practice the Strategy

Mrs. Brown's class had a pet parade.
3 students each brought a brown rabbit.
2 students each brought a white rabbit.
1 student brought a black rabbit. How many rabbits were in the parade?

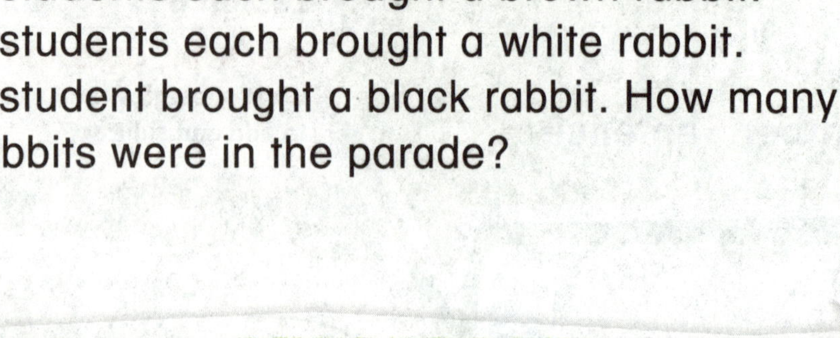

1 Understand Underline what you know. Circle what you need to find.

2 Plan How will I solve the problem?

3 Solve I will...

4 Check Is my answer reasonable? Explain.

Name ....

Mathematical PRACTICE

Apply the Strategy

Write a number sentence to solve.

1. There are 3 rabbits at a farm. The farmer buys 9 more. How many rabbits are there now?

_____ rabbits

2. Chung always picks 5 flowers at home after school. How many flowers will he have after 3 days?

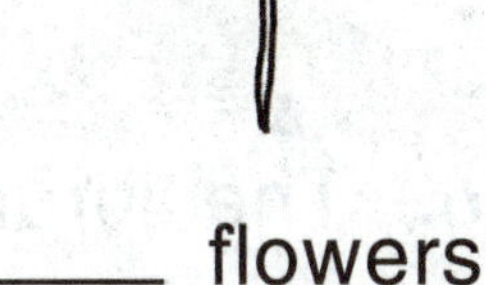

_____ flowers

3. Mrs. Lewis has 4 books for her reading group. She finds 2 more books. How many books does she have now?

_____ books

Review the Strategies

Choose a strategy
- Write a number sentence.
- Act it out.
- Draw a diagram.

4. Steve, Elena, and Ian are playing with toy planes. Steve gave 8 planes to Ian and 8 to Elena. He has 4 planes left. How many planes did Steve start with?

_______ planes

5. Arielle has 4 markers. Julia has 3 more markers than Arielle. How many markers do they have in all?

_______ markers

6. The library has 4 books about parrots and 3 books about swans. There is 1 less book about blue jays than swans. How many books about birds are there in all?

_______ books

Name

My Homework

Lesson 6
Problem Solving: Write a Number Sentence

Kim's mom is making blueberry pancakes.
She makes 4 pancakes in the first batch.
In the second batch she makes 5 more.
How many pancakes does Kim's mom make?

eHelp

1 Understand Underline what you know.
Circle what you need to find.

2 Plan How will I solve the problem?

3 Solve I will write a number sentence.

$4 + 5 = 9$

Kim's mom made 9 pancakes.

4 Check Is my answer reasonable?

Practice

Underline what you know. Circle what you need to find. Write a number sentence to solve.

1. Kit, Ruth, and Dean each eat 2 slices of pizza. How much pizza do they eat in all?

_____ + _____ + _____ = _____ pizza slices

2. Aki packs 2 suitcases with 9 sweaters in each. How many sweaters did he pack in all?

_____ + _____ = _____ sweaters

3. Kylee is making a pattern with 6 circles and 3 squares. How many shapes does she use in all?

_____ + _____ = _____ shapes

Test Practice

4. Which addition sentence is not correct?

$4 + 4 + 2 = 10$ ○

$4 + 3 + 3 = 10$ ○

$4 + 2 + 2 = 8$ 

$4 + 2 + 4 = 8$ 

Math at Home Have your child write a number sentence to solve this problem. There are two kittens eating. Three more come over to eat. How many kittens are there in all?

Name

Check My Progress

Vocabulary Check

Draw lines to match.

1. **add** — Both addends are the same.

2. **doubles** — 3 + 4 = 7

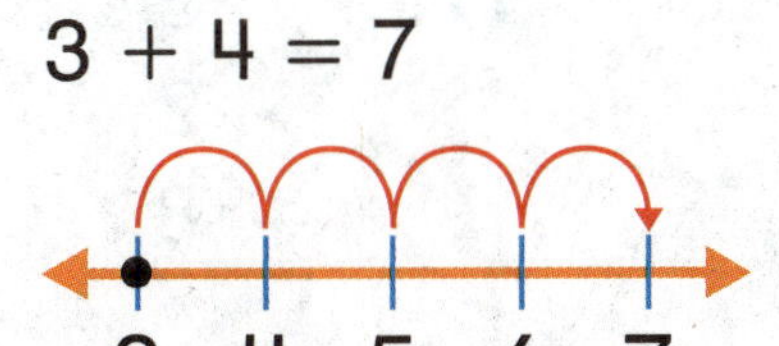

3. **count on** — The numbers you add in an addition number sentence.

4. **addends** — 6 + 8 = 14

Concept Check

Find each sum.

5. 7 + 0 = ______
   0 + 7 = ______

6. 3 + 6 = ______
   6 + 3 = ______

7. 2 + 5 = ______
   5 + 2 = ______

Add. Circle the doubles facts.

8. $0 + 3$

9. $1 + 1$

10. $3 + 4$

11. $9 + 9$

Make a ten to add.

12. $4 + 8$

13. $9 + 5$

14. $7 + 6$

15. $6 + 8$

Find each sum.

16. $1 + 4 + 6$

17. $5 + 2 + 8$

18. $7 + 1 + 2$

19. $8 + 8 + 2$

20. Lou saw 6 penguins at the zoo.
He also saw 3 flamingos and 2 pelicans.
How many birds did he see in all?

_____ birds

Test Practice

21. Ali's dog had 8 puppies. Luca's dog had the same number of puppies. How many puppies are there in all?

8 ○ 16 ○ 18 ○ 88 ○

Name ______________________

Count Back to Subtract

Lesson 7

ESSENTIAL QUESTION
What strategies can I use to add and subtract?

Explore and Explain

There are SO many to choose from!

Teacher Directions: Draw a blue box around number 9. Count back 3. Circle that number red. Draw a green box around number 19. Count back 2. Circle that number yellow. Draw a purple box around 14. Count back 1. Circle that number orange.

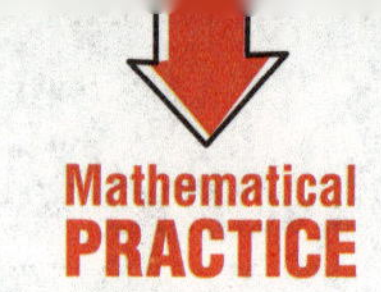

See and Show

Find 10 − 3. Use a number line. **Count back** to **subtract**.
The answer is the **difference**.

Start with the greater number.

No flowers on this page!
I want to go back.

$10 - 3 =$ 7
difference

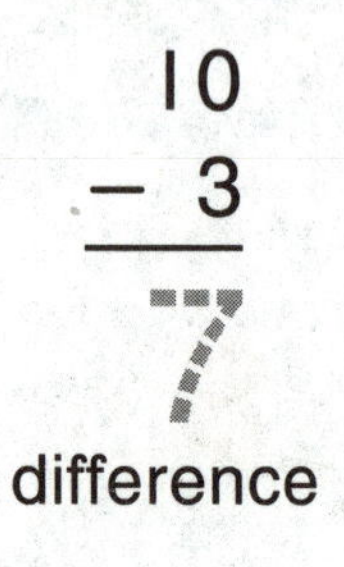

difference

Count back to subtract. Use the number line.

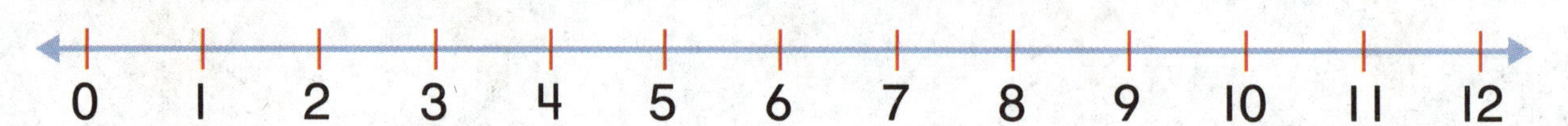

1. $7 - 3 =$ ______
2. $8 - 1 =$ ______
3. $9 - 2 =$ ______
4. $10 - 2 =$ ______
5. $8 - 2 =$ ______
6. ______ $= 7 - 1$

Talk Math Explain how you count back on a number line to subtract.

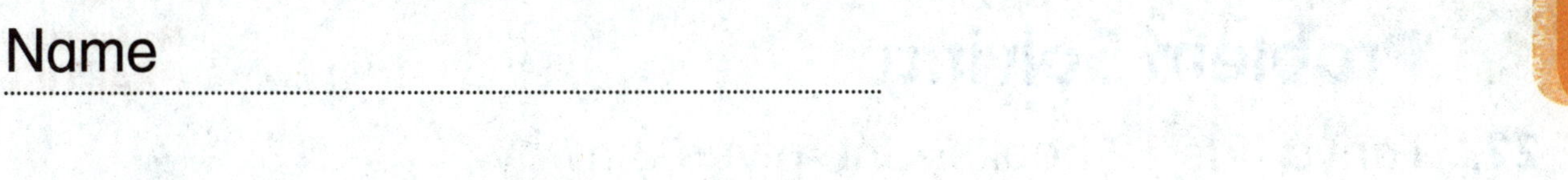

Name ______________________

On My Own

Count back to subtract. Use the number line.

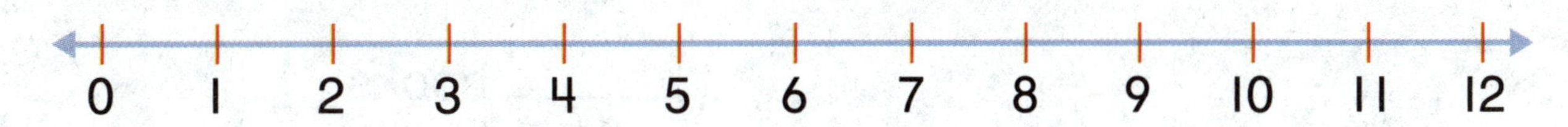

7. 3 − 2 = ______

8. ______ = 12 − 3

9. ______ = 8 − 3

10. 11 − 3 = ______

11. 9 − 3 = ______

12. 11 − 2 = ______

13. $\begin{array}{r} 9 \\ -\ 1 \\ \hline \end{array}$

14. $\begin{array}{r} 5 \\ -\ 3 \\ \hline \end{array}$

15. $\begin{array}{r} 10 \\ -\ 1 \\ \hline \end{array}$

16. $\begin{array}{r} 4 \\ -\ 2 \\ \hline \end{array}$

17. $\begin{array}{r} 12 \\ -\ 2 \\ \hline \end{array}$

18. $\begin{array}{r} 4 \\ -\ 3 \\ \hline \end{array}$

19. $\begin{array}{r} 8 \\ -\ 1 \\ \hline \end{array}$

20. $\begin{array}{r} 7 \\ -\ 2 \\ \hline \end{array}$

21. $\begin{array}{r} 5 \\ -\ 2 \\ \hline \end{array}$

Problem Solving

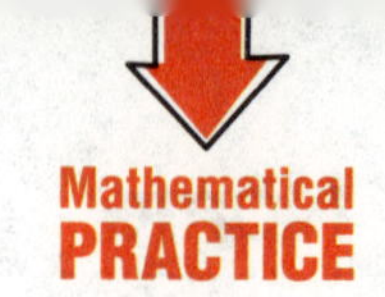

22. Tanya has 12 books. She gives 3 away. How many books does she have now?

_____ books

23. Hank washes 9 windows at the pet store. He washes 2 windows in the morning. How many windows does he wash in the afternoon?

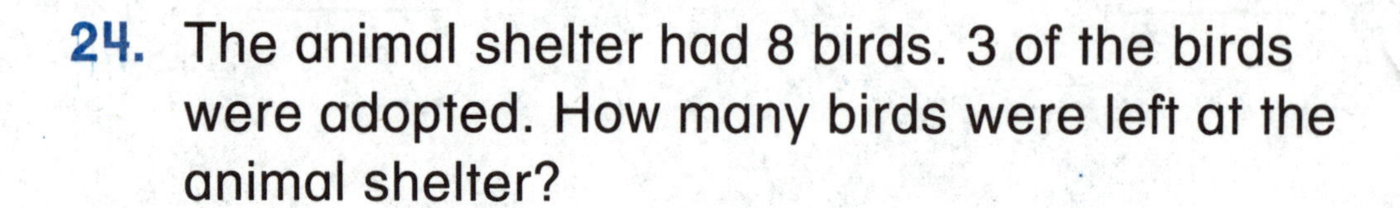

_____ windows

24. The animal shelter had 8 birds. 3 of the birds were adopted. How many birds were left at the animal shelter?

_____ birds

Explain how counting back can help you subtract.

____

____

____

Name ______________________

My Homework

Lesson 7
Count Back to Subtract

Homework Helper

eHelp

Need help? connectED.mcgraw-hill.com

Use the number line. Start at 9, count back to subtract.

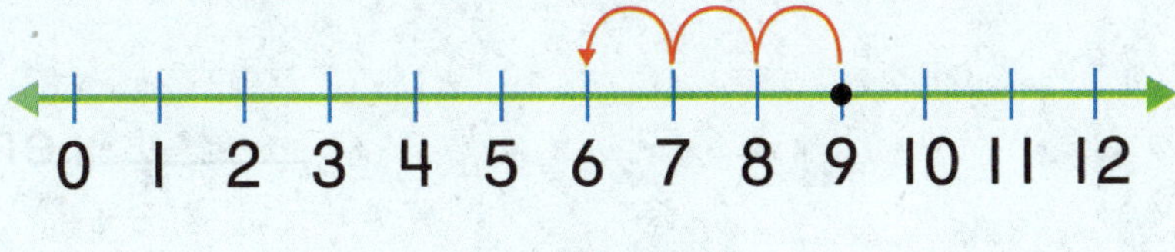

9 − 3 = 6

Practice

Count back to subtract. Use the number line above.

1. 12 − 4 = ______
2. 11 − 3 = ______
3. 8 − 3 = ______
4. 6 − 2 = ______
5. 6 − 3 = ______
6. 10 − 2 = ______

7. $\begin{array}{r} 12 \\ -\ 3 \\ \hline \end{array}$

8. $\begin{array}{r} 11 \\ -\ 1 \\ \hline \end{array}$

9. $\begin{array}{r} 8 \\ -\ 2 \\ \hline \end{array}$

10. A paper clip holder has 12 clips. Alex uses 3 paper clips. How many paper clips are left?

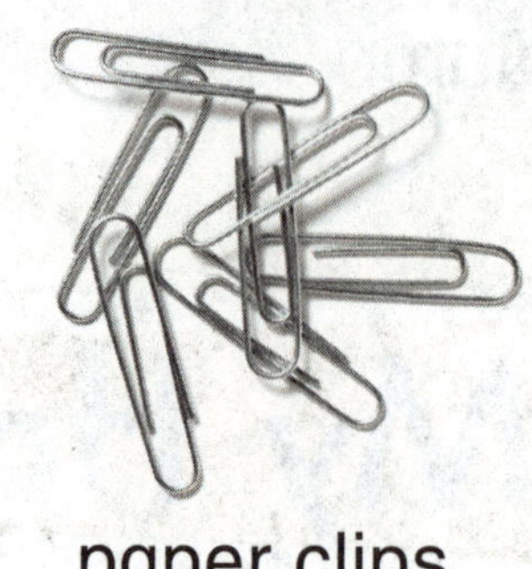

_______ paper clips

11. Marty buys 11 erasers. He uses 1 eraser. How many erasers does Marty have left?

_______ erasers

12. Jenny had 9 pencils in her desk. She let her friends borrow 2 of them. How many pencils does she have left?

_______ pencils

Vocabulary Check

Complete the sentence.

count back **sum** **difference**

13. You can ______________ to subtract.

14. The answer to a subtraction problem is called the ______________.

Math at Home Have your child count back to subtract. Say a number between 3 and 12. Have you child subtract 1, 2, or 3.

Name ______________________________

Subtract All and Subtract Zero

Lesson 8

ESSENTIAL QUESTION
What strategies can I use to add and subtract?

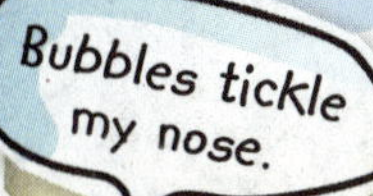

Explore and Explain

I'm drenched!

_____ − _____ = _____ dogs

Teacher Directions: Draw a picture and write a number sentence to solve the problem. Mary helped wash dogs at the pet store. There were 7 dogs to be washed. Mary washed 7 of them. How many dogs were left to be washed?

See and Show

You can subtract to find the difference.
Find how many frogs are left.

Subtract all.

5 − 5 = 0

0 frogs are left.

Subtract zero.

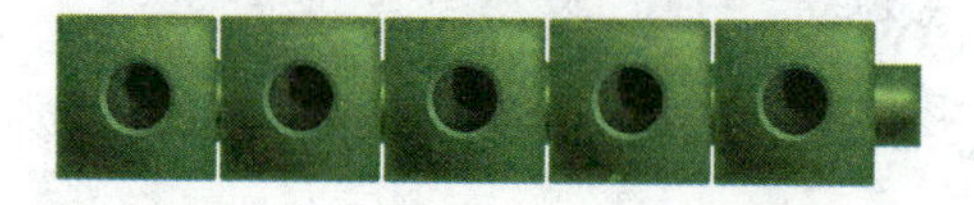

5 − 0 = 5

5 frogs are left.

Subtract.

1. 13 − 13 = ______

   13 − 0 = ______

2. ______ = 7 − 0

   7 − 7 = ______

3. 6 − 6 = ______

   ______ = 6 − 0

4. 8 − 0 = ______

   8 − 8 = ______

Talk Math Explain how you know 8 − 8 = 0 and 8 − 0 = 8.

Name

CCSS

On My Own

Subtract. Circle the problem if the difference is zero.

Helpful Hint
When you subtract zero from a number, the difference is the same as that number.

5. $9 - 3 =$ ______

6. ______ $= 8 - 1$

7. ______ $= 4 - 4$

8. $7 - 3 =$ ______

9. $8 - 8 =$ ______

10. $8 - 2 =$ ______

11. $6 - 0 =$ ______

12. $5 - 5 =$ ______

13. $3 - 2 =$ ______

14. $16 - 16 =$ ______

15. $7 - 2 =$ ______

16. $4 - 0 =$ ______

17. $9 - 0 =$ ______

18. $8 - 0 =$ ______

19. $9 - 9 =$ ______

20. $12 - 2 =$ ______

21. $11 - 6 =$ ______

22. $3 - 3 =$ ______

23. $7 - 1 =$ ______

24. $6 - 6 =$ ______

Problem Solving

25. 13 bees buzz near a flower. None fly away. How many bees are near the flower?

_____ bees

26. 4 sparrows are in the nest. They all fly away. How many sparrows are still in the nest?

_____ sparrows

27. Todd had 12 beads. He lost some of them. Now he has 7 beads. How many beads did Todd lose?

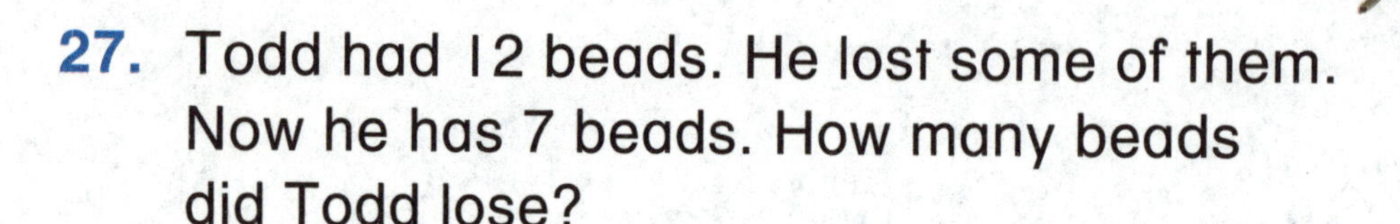

_____ beads

In a subtraction number sentence, which number is the difference?

____

____

____

Name ....

My Homework

Lesson 8
Subtract All and Subtract Zero

Homework Helper

Need help? connectED.mcgraw-hill.com

You can subtract all.

$6 - 6 = 0$

You can subtract zero.

$6 - 0 = 6$

Practice

Subtract.

1. $9 - 0$ = ____
2. $15 - 15$ = ____
3. $1 - 1$ = ____
4. $7 - 0$ = ____
5. $12 - 12$ = ____
6. $4 - 4$ = ____
7. $8 - 0$ = ____
8. $0 - 0$ = ____
9. $3 - 0 =$ ____
10. $17 - 17 =$ ____
11. $3 - 3 =$ ____
12. $2 - 0 =$ ____

13. 10 children play ball. After they finish, all 10 go back to class. How many children keep playing ball?

Recess rules!

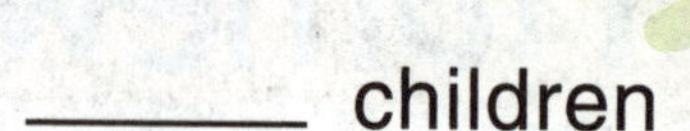

_____ children

14. 8 girls take a walk. When they reach the park, they all keep walking. How many girls are still taking a walk?

_____ girls

15. 18 children are at the store. All of them check out and go outside. How many are still shopping?

_____ children

Test Practice

16. Find the difference.

$19 - 19 =$

0	18	19	38
○	○	○	○

Math at Home Have your child use small objects to show 5 – 5 and 5 – 0.

Name ....

Use Doubles to Subtract

Lesson 9

ESSENTIAL QUESTION
What strategies can I use to add and subtract?

Explore and Explain

Hey! I'm seeing double!

______ cubes ______ cubes

Teacher Directions: Place a 6 cube train at the top of the page. Trace it. Separate the train into two equal parts. Draw the parts below the 6 cube train. Write how many cubes are in each part.

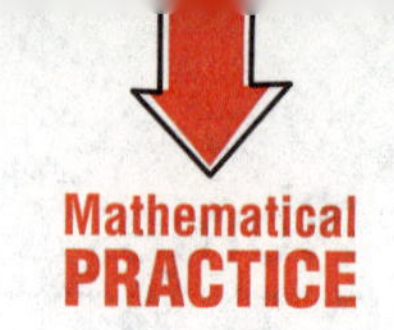

See and Show

You can use doubles facts to help you subtract.
Find 16 − 8.

If you know 8 + 8 = 16,

then 16 − 8 = 8.

Use cubes and doubles facts to subtract.

1. 

9 + ______ = 18

18 − 9 = ______

2. 

5 + ______ = 10

10 − 5 = ______

3. 6 + ______ = 12

12 − 6 = ______

4. 4 + ______ = 8

8 − 4 = ______

5. 7 + ______ = 14

14 − 7 = ______

6. 8 + ______ = 16

16 − 8 = ______

Talk Math Explain how you can use a doubles fact to subtract.

CCSS

Name

Helpful Hint
I know $5 + 5 = 10$,
so $10 - 5 = 5$.

On My Own

Subtract. Circle the doubles facts.

7. $9 - 3 =$ ______

8. $11 - 9 =$ ______

9. $6 -$ ______ $= 3$

10. $7 - 3 =$ ______

11. $18 - 9 =$ ______

12. $15 - 7 =$ ______

13. $16 - \square = 8$

14. $7 - 0$

15. $10 - \square = 5$

16. $9 - 8$

17. $6 - 2$

18. $8 - 3$

19. $14 - \square = 7$

20. $7 - 6$

21. $8 - \square = 4$

Problem Solving

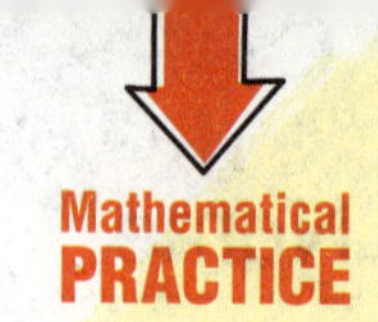

22. Fran and her grandmother pick 16 pumpkins. They use 8 pumpkins for pie. How many pumpkins are left?

_____ pumpkins

23. Delia bakes 18 cherry pies. She sells 9 pies at a farmers' market. How many pies does Delia have left?

_____ pies

24. 14 children were playing hide and go seek. 7 children were found. How many children are still hiding?

_____ children

Explain how you can use doubles facts to help you subtract.

____

____

____

Name ____________________

My Homework

Lesson 9

Use Doubles to Subtract

Homework Helper

Need help? connectED.mcgraw-hill.com

Use doubles facts to subtract. Find $12 - 6$.

If you know $6 + 6 = 12$,

then you also know $12 - 6 = 6$.

Practice

Subtract. Circle the doubles facts.

1. $18 - 9 =$ ______
2. $6 - 2 =$ ______
3. $8 - 2 =$ ______
4. $5 - 3 =$ ______
5. $16 - 8 =$ ______
6. $8 - 3 =$ ______
7. $6 - 3 =$ ______
8. $8 - 8 =$ ______
9. $\begin{array}{r} 14 \\ -\ 7 \\ \hline \end{array}$
10. $\begin{array}{r} 9 \\ -\ 1 \\ \hline \end{array}$
11. $\begin{array}{r} 10 \\ -\ 4 \\ \hline \end{array}$

Subtract. Circle the doubles facts.

12. $14 - 6 =$ ______

13. $18 - 8 =$ ______

14. $18 - 9 =$ ______

15. $15 - 7 =$ ______

16. Anita checks out 14 library books. She reads 7 of the books. How many books does she still have to read?

______ books

17. Lindsay has 16 cupcakes to hand out. She has given 8 to the girls in her class. How many does she have left?

______ cupcakes

Test Practice

18. Mark the number sentence that uses doubles.

$4 - 4 = 0$  $8 - 4 = 4$  $8 - 3 = 5$  $8 - 8 = 0$ ○

Math at Home Call out doubles facts. Have your child name the subtraction problem for each double.

Name ....

Check My Progress

Vocabulary Check

Complete each sentence.

count back **subtract** **difference**

1. You can ________________ by taking a number away from a greater number.

2. The number you have left after you subtract is called the ________________.

3. You can ________________ on a number line to help you subtract.

Concept Check

Count back to subtract. Use the number line to help.

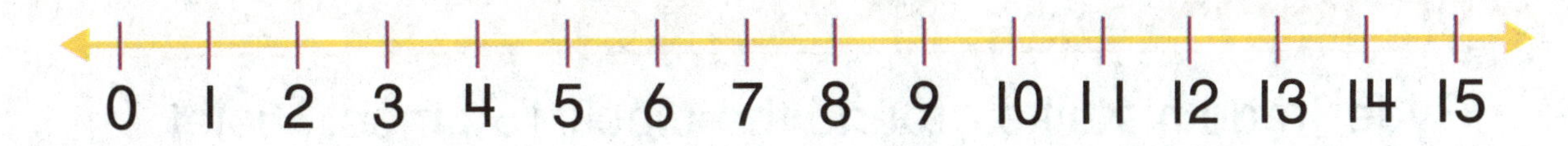

4. $13 - 3 =$ ________

5. $15 - 2 =$ ________

6. $7 - 3 =$ ________

7. $14 - 3 =$ ________

8. $10 - 1 =$ ________

9. $12 - 8 =$ ________

Subtract.

10. $\begin{array}{r} 6 \\ -\ 0 \\ \hline \end{array}$

11. $\begin{array}{r} 8 \\ -\ 8 \\ \hline \end{array}$

12. $\begin{array}{r} 9 \\ -\ 0 \\ \hline \end{array}$

13. $\begin{array}{r} 4 \\ -\ 4 \\ \hline \end{array}$

Use doubles or near doubles facts to subtract.

14. $\begin{array}{r} 16 \\ -\ 8 \\ \hline \end{array}$

15. $\begin{array}{r} 7 \\ -\ 3 \\ \hline \end{array}$

16. $\begin{array}{r} 10 \\ -\ 4 \\ \hline \end{array}$

17. $\begin{array}{r} 18 \\ -\ 9 \\ \hline \end{array}$

18. $\begin{array}{r} 14 \\ -\ 7 \\ \hline \end{array}$

19. $\begin{array}{r} 8 \\ -\ 5 \\ \hline \end{array}$

20. $\begin{array}{r} 6 \\ -\ 3 \\ \hline \end{array}$

21. $\begin{array}{r} 9 \\ -\ 4 \\ \hline \end{array}$

22. Laine counts 18 fireflies. She counts 9 fireflies in the front yard. How many fireflies does Laine count in the backyard?

_______ fireflies

Test Practice

23. Ryder rode his bike around the block 16 times. Violet rode half as many times. How many times did Violet ride her bike around the block?

16	10	8	6
○	○	○	○

Name ....

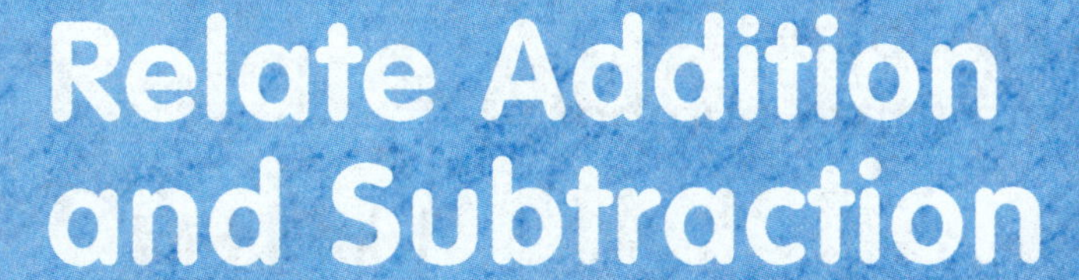

Relate Addition and Subtraction

Lesson 10

ESSENTIAL QUESTION
What strategies can I use to add and subtract?

Explore and Explain

Part	Part
Whole	

Did those counters move?

_____ + _____ = _____

_____ − _____ = _____

Teacher Directions: Place 4 red counters in one part. Place 3 yellow counters in another part. Move the counters to the whole to show the total. Write the addition sentence. Move the yellow counters back to the part. Write the subtraction sentence shown.

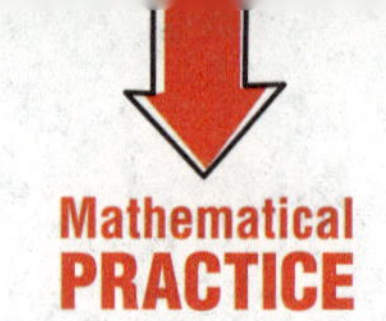

See and Show

You can use addition facts to subtract.
Related facts have the same three numbers.

Helpful Hint
Addition and subtraction are opposite or inverse operations.

5 + 4 = 9

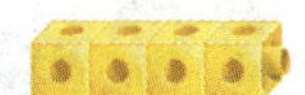

9 − 4 = 5

9 − 5 = 4

Use the addition facts to subtract.

1. 

6 + 7 = ______

13 − 6 = ______

13 − 7 = ______

2. 

5 + 7 = ______

12 − 7 = ______

12 − 5 = ______

3. 5 + 6 = ______

11 − 6 = ______

11 − 5 = ______

4. 7 + 4 = ______

11 − 4 = ______

11 − 7 = ______

Talk Math Explain how addition and subtraction are related.

CCSS

Name ....

On My Own

Use addition facts to subtract.

5. $3 + 9 = \_\_\_$

$12 - 9 = \_\_\_$

$12 - 3 = \_\_\_$

6. $4 + 7 = \_\_\_$

$11 - 4 = \_\_\_$

$11 - 7 = \_\_\_$

7. $\begin{array}{r} 8 \\ +\ 9 \\ \hline \end{array}$ $\begin{array}{r} 17 \\ -\ 8 \\ \hline \end{array}$

8. $\begin{array}{r} 9 \\ +\ 6 \\ \hline \end{array}$ $\begin{array}{r} 15 \\ -\ 9 \\ \hline \end{array}$

9. $\begin{array}{r} 9 \\ +\ 5 \\ \hline \end{array}$ $\begin{array}{r} 14 \\ -\ 5 \\ \hline \end{array}$

10. $\begin{array}{r} 8 \\ +\ 4 \\ \hline \end{array}$ $\begin{array}{r} 12 \\ -\ 8 \\ \hline \end{array}$

11. $\begin{array}{r} 8 \\ +\ 7 \\ \hline \end{array}$ $\begin{array}{r} 15 \\ -\ 7 \\ \hline \end{array}$

12. $\begin{array}{r} 8 \\ +\ 5 \\ \hline \end{array}$ $\begin{array}{r} 13 \\ -\ 8 \\ \hline \end{array}$

13. $8 + 9 = \_\_\_$

$17 - 8 = \_\_\_$

14. $6 + 8 = \_\_\_$

$14 - 6 = \_\_\_$

15. $6 + 7 = \_\_\_$

$13 - 6 = \_\_\_$

16. $3 + 8 = \_\_\_$

$11 - 8 = \_\_\_$

17. $8 + 8 = \_\_\_$

$16 - 8 = \_\_\_$

18. $5 + 5 = \_\_\_$

$10 - 5 = \_\_\_$

Problem Solving

19. The tennis team has 16 players. 8 players leave the team. How many players are still on the team?

_____ players

Write a related fact. _____ ◯ _____ ◯ _____

I'm lucky I found this log!

20. 10 turtles are on a log. 2 turtles jump in the water and swim away. How many turtles are left on the log?

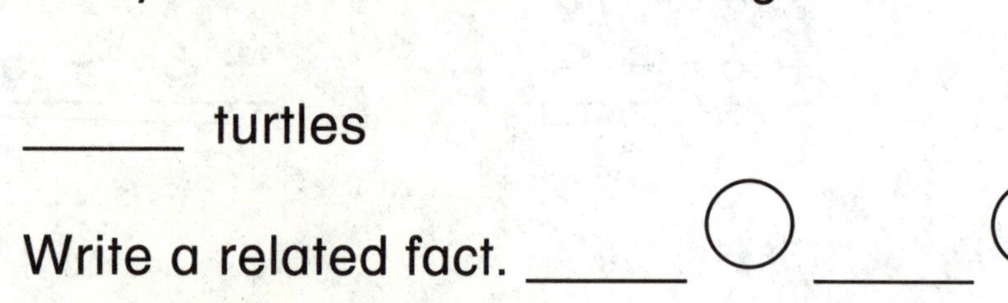

_____ turtles

Write a related fact. _____ ◯ _____ ◯ _____

21. 9 birds are flying. 3 birds fly away. How many birds are left flying?

_____ birds

Write a related fact. _____ ◯ _____ ◯ _____

Write an example of related facts. Use the numbers 4, 6, and 10.

____

____

____

My Homework

Lesson 10

Relate Addition and Subtraction

Homework Helper

Need help? connectED.mcgraw-hill.com

You can use addition facts to subtract.
If you know $6 + 3 = 9$, then you also know that $9 - 3 = 6$ and $9 - 6 = 3$.

$6 + 3 = 9$

$9 - 3 = 6$

$9 - 6 = 3$

Practice

Use addition facts to subtract.

1. $7 + 5 =$ ______

   $12 - 5 =$ ______

2. $6 + 9 =$ ______

   $15 - 9 =$ ______

3. $8 + 5 =$ ______

   $13 - 5 =$ ______

4. $4 + 7 =$ ______

   $11 - 7 =$ ______

Use addition facts to subtract.

5. $\begin{array}{r} 7 \\ +\ 3 \\ \hline \end{array}$ $\begin{array}{r} 10 \\ -\ 3 \\ \hline \end{array}$

6. $\begin{array}{r} 4 \\ +\ 0 \\ \hline \end{array}$ $\begin{array}{r} 4 \\ -\ 0 \\ \hline \end{array}$

7. $\begin{array}{r} 2 \\ +\ 8 \\ \hline \end{array}$ $\begin{array}{r} 10 \\ -\ 2 \\ \hline \end{array}$

8. Megan sent 12 postcards last week. This week she sent 4 postcards. How many more postcards did she send last week?

_______ postcards

Vocabulary Check 

9. **Using 2, 4, and 6, write an example of related facts.**

_____ ◯ _____ ◯ _____

_____ ◯ _____ ◯ _____

_____ ◯ _____ ◯ _____

_____ ◯ _____ ◯ _____

Math at Home Ask your child to show you an addition sentence with spoons and relate it to subtraction.

Name ....

Missing Addends

Lesson 11

ESSENTIAL QUESTION
What strategies can I use to add and subtract?

Explore and Explain

4 + ☐ = 9

2 + ☐ = 8

6 + ☐ = 11

3 + ☐ = 12

4 + ☐ = 8

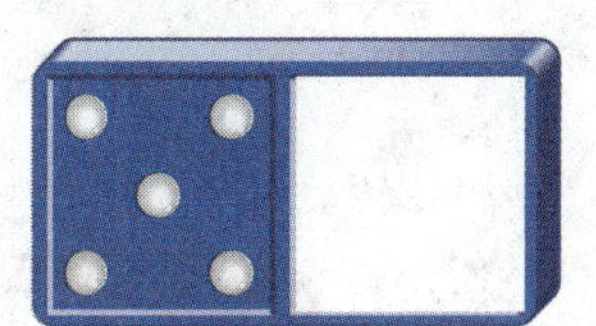

5 + ☐ = 10

If I were a domino, I'd be worth 4!

Teacher Directions: How many dots do you see? How many dots should there be in all? Draw more dots on the blank side of each domino until you have that number of dots in all. Complete the number sentence.

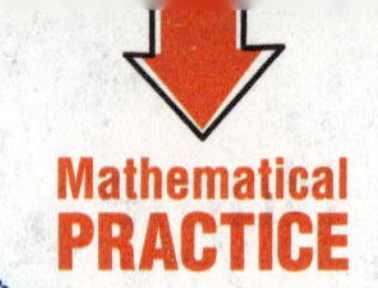

See and Show

You can use a related fact to help you find a **missing addend**.

9 + ☐ = 15

Think 15 − 9 = 6.

9 + 6 = 15

6 is the missing addend.

Find the missing addend. Draw that many dots on the domino.

1. 

8 + ☐ = 14

Think: 14 − 8 = ☐

2. 

5 + ☐ = 11

Think: 11 − 5 = ☐

3. 

☐ + 4 = 8

Think: 8 − 4 = ☐

4. 

7 + ☐ = 15

Think: 15 − 7 = ☐

Talk Math How do you find the missing addend in 5 + ☐ = 13?

Name ....

Helpful Hint
To find the missing addend, use related facts.

CCSS

On My Own

Find the missing addend.

5. 9 + ☐ = 12

12 − 9 = ☐

6. 7 + ☐ = 8

8 − 7 = ☐

7. ☐ + 6 = 12

12 − 6 = ☐

8. ☐ + 2 = 7

7 − 2 = ☐

9. 7 + ☐ = 16

16 − 7 = ☐

10. ☐ + 7 = 14

14 − 7 = ☐

11. 9 + ☐ = 9

9 − 9 = ☐

12. 6 + ☐ = 13

13 − 6 = ☐

13. 4 + ☐ = 12

12 − 4 = ☐

14. 5 + ☐ = 12

12 − 5 = ☐

Problem Solving

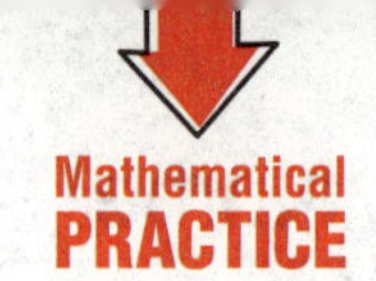

15. There were 14 seals on a rock. Some seals jumped into the water. 9 seals were left on the rock. How many seals jumped into the water?

______ seals

16. Anna buys 7 plants. She wants 12 plants. How many more plants does Anna need?

______ plants

17. J.J. wants 16 fish in his aquarium. He has 6 fish now. How many more fish does J.J. need?

______ fish

HOT Problem Julie found a missing addend like this:

$7 + \square = 9 \qquad 9 + 7 = 16$

Tell what she did wrong. Make it right.

____

____

____

Name ....

My Homework

Lesson 11
Missing Addends

Homework Helper

Need help? connectED.mcgraw-hill.com

Use related facts to find missing addends.

$7 + \boxed{8} = 15$ $15 - 7 = \boxed{8}$

Practice

Find the missing addend.

1. $8 + \boxed{} = 12$

$12 - 8 = \boxed{}$

2. $\boxed{} + 6 = 15$

$15 - 6 = \boxed{}$

3. $8 + \boxed{} = 13$

$13 - 8 = \boxed{}$

4. $\boxed{} + 7 = 11$

$11 - 7 = \boxed{}$

5. $8 + \boxed{} = 17$

$17 - 8 = \boxed{}$

6. $9 + \boxed{} = 12$

$12 - 9 = \boxed{}$

7. David and his friends are flying 9 kites. Some kites get stuck in a tree. 7 kites are still flying. How many kites are in the tree?

$7 + \square = 9$ ____ kites

8. The Girl Scouts put some boats in the pond. They leave 9 boats on land. There are 15 boats altogether. How many boats did the scouts put into the pond?

$9 + \square = 15$ ____ boats

9. 7 boys were riding their bikes. Some of the boys went home. 3 boys are still riding their bikes. How many boys went home?

$3 + \square = 7$ ____ boys

Vocabulary Check

Complete the sentence.

missing addend **sum**

10. When a number sentence only has one addend, you must find the ________________ to solve it.

Math at Home Ask your child to tell you the subtraction fact that will help him or her find the missing addend in $7 + \square = 15$.

Name ....

Fact Families

Lesson 12

ESSENTIAL QUESTION

What strategies can I use to add and subtract?

Explore and Explain

I think I see my cousin!

___ + ___ = ___

___ + ___ = ___

___ − ___ = ___

___ − ___ = ___

___ + ___ = ___

___ + ___ = ___

___ − ___ = ___

___ − ___ = ___

Teacher Directions: Roll a red number cube two times to find two addends. Write the numbers at the bottom of the triangle. Write the sum at the top of the triangle. Use the numbers to write related number sentences. Repeat for the other triangle.

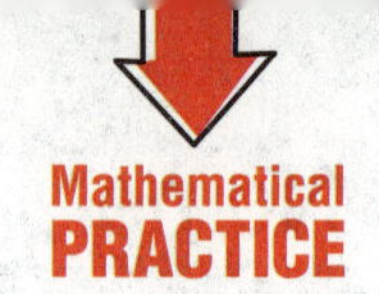

See and Show

A **fact family** is a set of related facts using the same three numbers. The numbers in this fact family are 12, 9, and 3.

9 + 3 = 12

3 + 9 = 12

12 − 9 = 3

12 − 3 = 9

Helpful Hint
Use related facts to complete fact families.

Complete each fact family.

1.

9 + 7 = ______

7 + 9 = ______

16 − 9 = ______

16 − 7 = ______

2.

9 + 8 = ______

______ + ______ = ______

17 − 9 = ______

______ − ______ = ______

Talk Math What are the related facts in the fact family 9, 9, 18? Explain.

Name ____________________

On My Own

Complete each fact family.

> **Helpful Hint**
> Each fact in a fact family uses the same three numbers.

3.

___ + ___ = ___

___ + ___ = ___

___ − ___ = ___

___ − ___ = ___

4.

___ + ___ = ___

___ + ___ = ___

___ − ___ = ___

___ − ___ = ___

5.

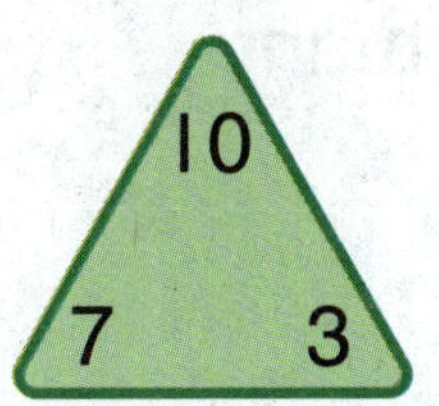

___ + ___ = ___

___ + ___ = ___

___ − ___ = ___

___ − ___ = ___

6.

___ + ___ = ___

___ + ___ = ___

___ − ___ = ___

___ − ___ = ___

Problem Solving

Solve.

7. Linda has 7 fish. She has 5 crabs. How many animals does Linda have?

_____ animals

8. There are 8 apples in one basket. The second basket has one more than the first. How many apples are there in all?

_____ apples

9. 11 children were on the playground. 3 more children came to the playground. How many children are on the playground now?

_____ children

HOT Problem Make a fact family. Use at least one two-digit number.

☐ + ☐ = ☐ ☐ − ☐ = ☐

☐ + ☐ = ☐ ☐ − ☐ = ☐

Name ....

My Homework

Lesson 12
Fact Families

Homework Helper

Need help? connectED.mcgraw-hill.com

Use related facts to complete a fact family.

$6 + 4 = 10$
$4 + 6 = 10$
$10 - 4 = 6$
$10 - 6 = 4$

Helpful Hint
Each fact in a fact family uses the same three numbers.

Practice

Complete each fact family.

1.

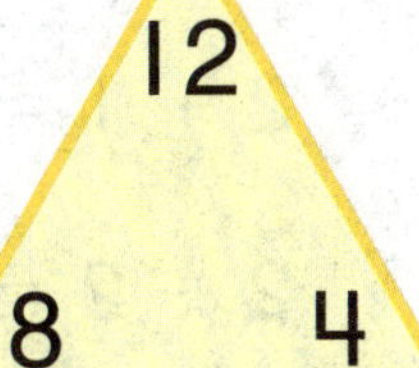

_____ + _____ = _____

_____ + _____ = _____

_____ − _____ = _____

_____ − _____ = _____

2.

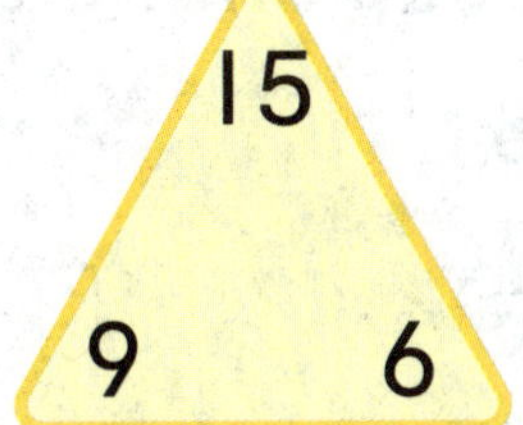

_____ + _____ = _____

_____ + _____ = _____

_____ − _____ = _____

_____ − _____ = _____

3. There are 8 apples on one tree.
There are 7 apples on another tree.
How many apples are there in all?

_______ apples

4. Sam has 7 toy trucks in one box. He has one less than that number in another box. How many toy trucks does he have in all?

_______ toy trucks

Vocabulary Check

Circle the correct answer.

5. **Which group shows a fact family?**

2 + 3 = 5	1 + 4 = 5	2 + 1 = 3
3 + 2 = 5	1 + 5 = 6	2 + 2 = 4
5 − 3 = 2	1 + 7 = 8	2 + 3 = 5
5 − 2 = 3	1 + 8 = 9	2 + 4 = 6

Math at Home Have your child write the fact family for the numbers 6, 7, and 13.

Name

Two-Step Word Problems

Lesson 13

ESSENTIAL QUESTION
What strategies can I use to add and subtract?

Explore and Explain

_____ dolphins

Teacher Directions: Use counters to solve the problem. Trace the counters to show your work. There is a group of 12 dolphins following a boat. 3 more dolphins join the group. Then 6 of the dolphins swim away. How many dolphins are still following the boat?

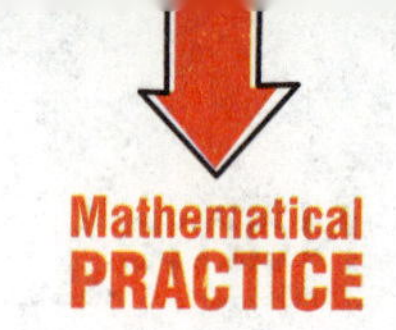

See and Show

Some word problems take two steps to solve.
Tim's class was having a pizza party. Tim brought 3 pizzas. Jenni brought 2 pizzas. The class ate 4 pizzas. How many pizzas do they have left?

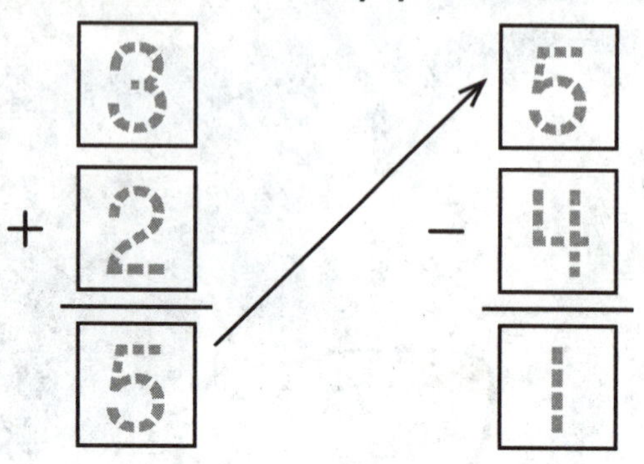

1 pizza

Solve each two-step word problem.

1. Marla bought 3 books on Monday and 5 books on Wednesday. She returned 4 books today. How many books does Marla have now?

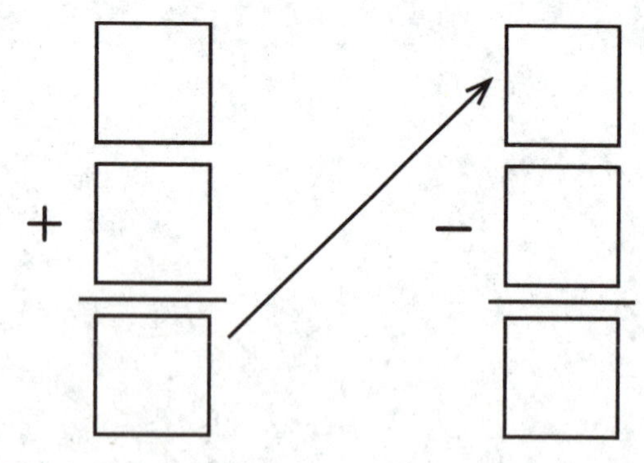

_____ books

2. Carla collected 6 leaves but she lost 3 of them. She has 9 leaves at home. How many leaves does Carla have in all?

_____ leaves

Talk Math Explain how to solve a two-step word problem.

Name

On My Own

Solve each two-step word problem.

3. Candice has 6 pencils. She gives 2 away. Then she gets 2 pencils out of her bag. How many pencils does Candice have now?

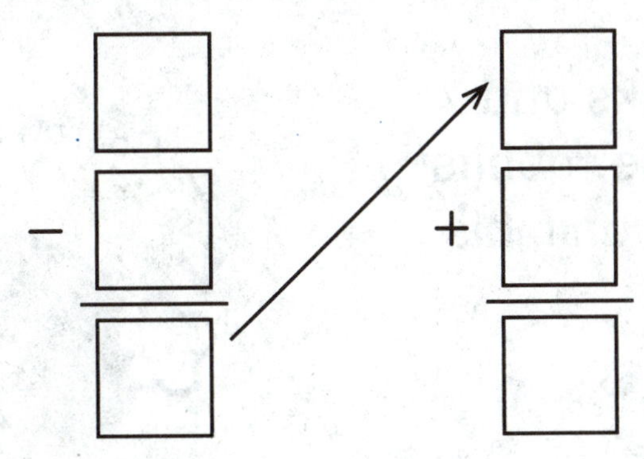

_______ pencils

4. 8 birds land on a tree. 5 of the birds fly to another tree. Then 2 more fly away. How many birds are left in the tree?

_______ birds

5. Suzie saw 9 ladybugs on her porch. 4 flew away. Then 3 more ladybugs came. How many ladybugs are on the porch now?

_______ ladybugs

Problem Solving

6. 5 seashells were lying on the beach. 8 more shells washed up onto the beach. Sammy picked up 4 shells. How many shells were left on the beach?

_____ shells

7. David counted 6 chipmunks and 7 rabbits in his yard. 5 of the rabbits hopped away. How many animals can David see now?

_____ animals

8. There were 5 toads in the pond. 3 more toads jumped in. 2 toads jumped out of the pond. How many toads are in the pond now?

_____ toads

HOT Problem Write a two-step word problem using the number sentences $8 + 2 = 10$ and $10 - 4 = 6$.

____

____

____

Name ______________________

My Homework

Lesson 13
Two-Step Word Problems

Homework Helper

Need help? connectED.mcgraw-hill.com

Tony bought 3 items for lunch. Delia bought 3 items for lunch. Sally had 4 items for lunch. How many items did they have in all?

$$\begin{array}{r} 3 \\ + \ 3 \\ \hline 6 \end{array} \qquad \begin{array}{r} 6 \\ + \ 4 \\ \hline 10 \end{array}$$

items

Practice

Solve each two-step word problem.

1. Camden has 3 marbles in his pocket. He finds 8 more. He gives 4 marbles to Jen. How many marbles does Camden have now?

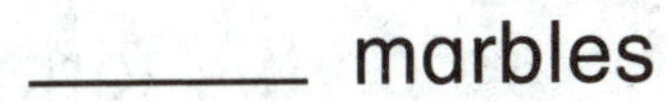
_______ marbles

2. Alexis collected 6 rocks. She already had 9 rocks. She lost 3 rocks. How many rocks does Alexis have now?

_______ rocks

Solve each two-step word problem.

3. 9 geese landed in the yard. 8 of the geese flew away. Then 3 more geese landed in the yard. How many geese are in the yard now?

_____ geese

4. Sheila had 15 cards to mail to her friends. She forgot to mail 2 of them. 6 of them got lost in the mail. How many cards were delivered?

_____ cards

5. Gia got 4 books from the library yesterday. She got 5 books today. She returned 3 books. How many library books does Gia have now?

_____ books

Test Practice

6. There are 7 horses on the farm. There are also 9 pigs. 2 of the pigs were sent to another farm. How many animals are on the farm now?

18	14	7	5
○	○	○	○

Math at Home Tell a two-step number story to your child. Have him or her use pennies to show you the steps they use to solve it.

Name ....

My Review

Chapter 1
Apply Addition and Subtraction Concepts

Vocabulary Check

add	**addend**	**count back**	**count on**
fact family	**related facts**	**subtract**	

Write the correct word in the blank.
Use words from the box.

1. When you take one part away from a whole, you ________________.

2. Addition and subtraction facts that use the same three numbers can be called a ________________ or ________________.

3. The two numbers you add together in an addition sentence are called ________________.

4. You can ________________ to subtract 2 from a number.

5. To ________________, you join two parts together.

6. You can ________________ to add 2 to a number.

Concept Check 

Add.

7. $\begin{array}{r} 9 \\ +\ 3 \\ \hline \end{array}$

8. $\begin{array}{r} 8 \\ +\ 5 \\ \hline \end{array}$

9. $\begin{array}{r} 9 \\ +\ 6 \\ \hline \end{array}$

10. $\begin{array}{r} 7 \\ +\ 7 \\ \hline \end{array}$

Find each sum.

11. $6 + 5 + 4 =$ ______

12. $2 + 2 + 8 =$ ______

Use addition facts to subtract.

13. $7 + 8 =$ ______

$15 - 7 =$ ______

14. $6 + 7 =$ ______

$13 - 7 =$ ______

Complete the fact family.

15. 

______ + ______ = ______

______ + ______ = ______

______ − ______ = ______

______ − ______ = ______

Name ....

Problem Solving

16. Dotty has 8 red flowers and 2 pink flowers. She gives 5 flowers to her mother. How many flowers does Dotty have left?

_______ flowers

17. Oscar finds 6 leaves in the morning. He finds 2 more leaves after lunch. He already has 8 leaves. How many leaves does he have in all?

_______ leaves

18. Shani walks the dogs in her neighborhood. One neighbor has 4 dogs. Another neighbor has 3 dogs. How many dogs does she walk?

_______ dogs

Test Practice

19. Rod throws 5 baseballs. Ty throws the same number of baseballs as Rod. How many baseballs did they throw in all?

0	5	10	15
○	○	○	○

Reflect

Chapter 1

Answering the Essential Question

Show different strategies to add and subtract.

Make a 10.

5 + 9 = ____

10 + ____ = ____

Use doubles to add and subtract.

7 + 7 = ____

14 − 7 = ____

ESSENTIAL QUESTION

What strategies can I use to add and subtract?

Complete the fact family.

2 + 5 = 7

____ + ____ = ____

____ − ____ = ____

____ − ____ = ____

Find the missing addend.

3 + ____ = 9

9 − 3 = ____

See, you can do it!

Chapter

2 Number Patterns

ESSENTIAL QUESTION

How can equal groups help me add?

We're Going to the Desert

Watch a video!

Watch

My Common Core State Standards

Operations and Algebraic Thinking

2.OA.1 Use addition and subtraction within 100 to solve one-and two-step word problems involving situations of adding to, taking from, putting together, taking apart, and comparing, with unknowns in all positions.

2.OA.2 Fluently add and subtract within 20 using mental strategies. By end of Grade 2, know from memory all sums of two one-digit numbers.

2.OA.3 Determine whether a group of objects (up to 20) has an odd or even number of members.

2.OA.4 Use addition to find the total number of objects arranged in rectangular arrays with up to 5 rows and up to 5 columns; write an equation to express the total as a sum of equal addends.

Number and Operations in Base Ten *This chapter also addresses this standard:*

2.NBT.2 Count within 1000; skip-count by 5s, 10s, and 100s.

1. Make sense of problems and persevere in solving them.
2. Reason abstractly and quantitatively.
3. Construct viable arguments and critique the reasoning of others.
4. Model with mathematics.
5. Use appropriate tools strategically.
6. Attend to precision.
7. Look for and make use of structure.
8. Look for and express regularity in repeated reasoning.

= focused on in this chapter

Name

Am I Ready?

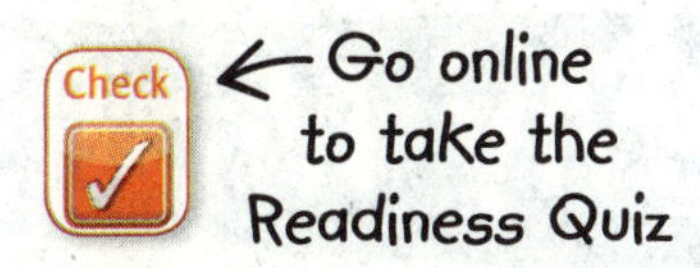

1. Write the missing numbers.

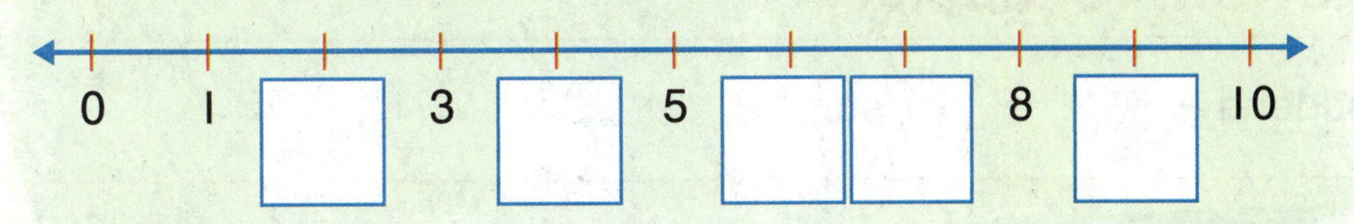

Write the missing numbers.

2. _______, 2, _______, _______, 5, 6, _______

3. 11, _______, _______, 14, _______, 16, _______

4. 10, 20, _______, 40, _______, 60, _______

Add.

5. $5 + 5 =$ _______

6. $2 + 2 + 2 =$ _______

7. Alisa has 3 red cups, 3 blue cups, and 3 pink cups. How many cups does she have in all? _______

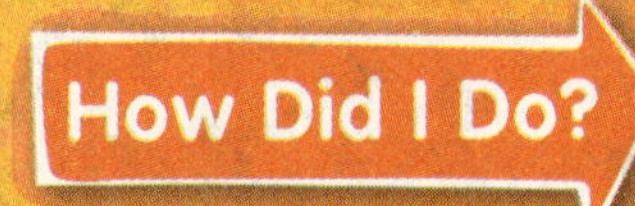

Shade the boxes to show the problems you answered correctly.

1	2	3	4	5	6	7

Name ....

My Math Words

Review Vocabulary

addend sum

Write a review word in each box to describe the parts of each addition number sentence.

7	+	9	=	16
☐		☐		☐

13	=	6	+	7
☐		☐		☐

5	+	8	=	13
☐		☐		☐

My Vocabulary Cards

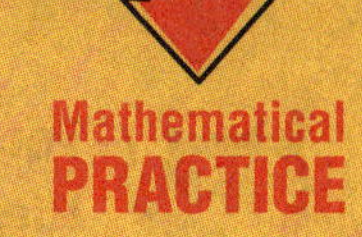

Lesson 2-5

array

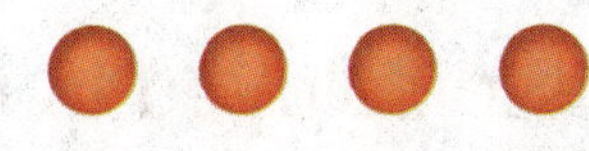
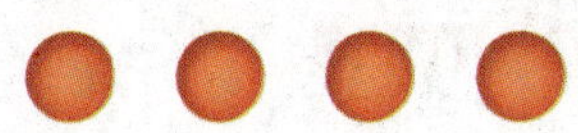

2 groups of 4

Lesson 2-2

equal groups

3 equal groups

Lesson 2-6

even

1, 3, 5,
17, 45, 89

2, 4, 6,
24, 52, 60

Lesson 2-6

odd

1, 3, 5,
17, 45, 89

2, 4, 6,
24, 52, 60

Lesson 2-4

repeated addition

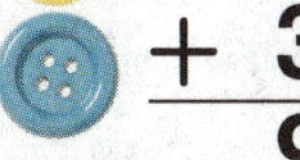

$$\begin{array}{r} 3 \\ 3 \\ +\ 3 \\ \hline 9 \end{array}$$

Lesson 2-1

skip count

5, 10, 15, 20, 25

+5 +5 +5 +5

Teacher Directions:
Ideas for Use

- Group 2 or 3 common words. Add a word that is unrelated to the group. Ask another student to name the unrelated word.
- Arrange the cards in alphabetical order.

Each group has the same number of objects.

Objects displayed in rows and columns.

An odd number of objects has 1 left over when counted by 2s. Numbers that end with 1, 3, 5, 7, or 9.

An even number of objects can be counted by 2s. Numbers that end with 0, 2, 4, 6, 8.

Count objects in equal groups of two or more.

To use the same addend over and over.

FOLDABLES® Follow the steps on the back to make your Foldable.

____ rows of ____

$3 + 3 =$ ____

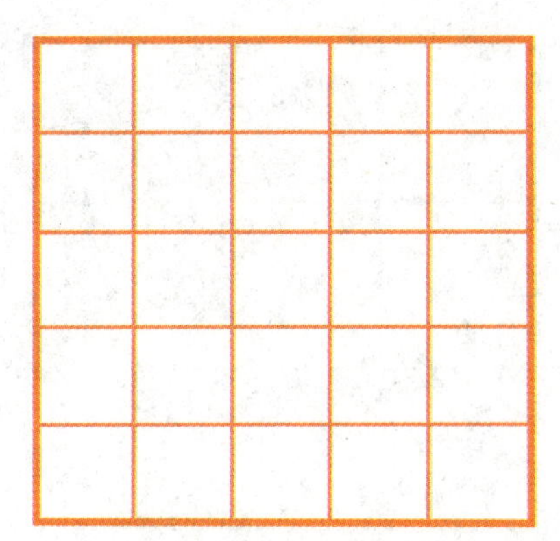

____ rows of ____

________________ = ____

Repeated Addition with

Arrays

1

2

3

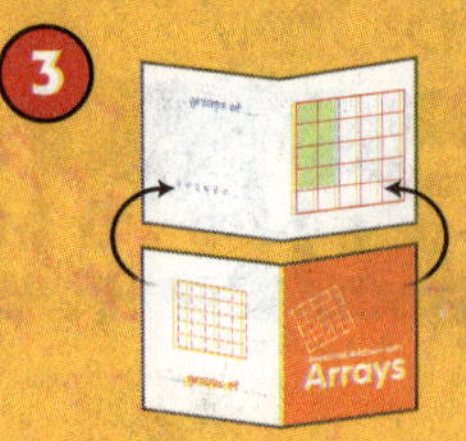

4

5 rows of 4

______________ = ______

______ rows of ______

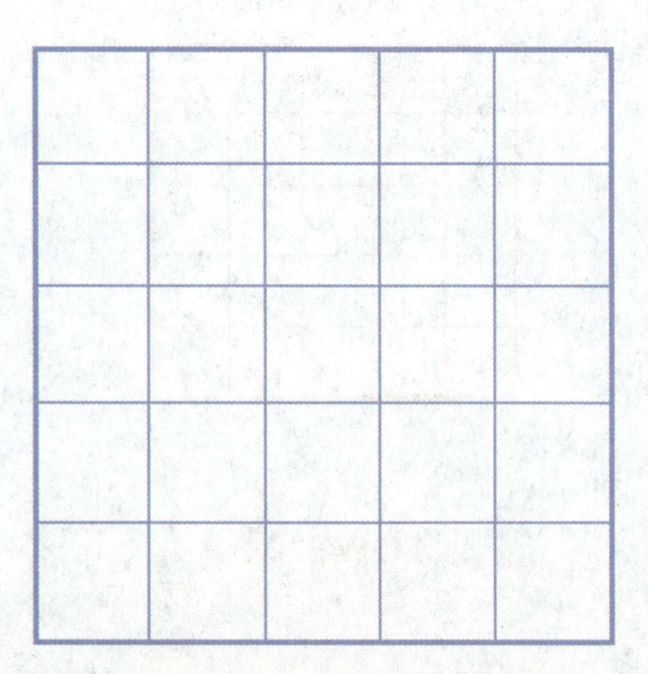

______________ = ______

Name ....

Skip Count On a Hundred Chart

Lesson 1

ESSENTIAL QUESTION
How can equal groups help me add?

Explore and Explain

1		3		5		7		9	
11		13		15		17		19	
21		23		25		27		29	
31		33		35		37		39	
41		43		45		47		49	
51		53		55		57		59	
61		63		65		67		69	
71		73		75		77		79	
81		83		85		87		89	
91		93		95		97		99	

Teacher Directions: Count by 2s. Write the missing numbers. Count by 5s. Color the numbers yellow. Count by 10s. Circle those numbers red.

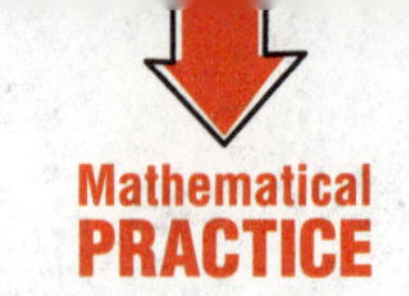

See and Show

Helpful Hint
Skip counting on a hundred chart creates patterns.

Use patterns to count. You can **skip count** by 2s, 5s, and 10s.

1	2	3	4	5	6	7	8	9	10
11	12	13	14	15	16	17	18	19	20
21	22	23	24	25	26	27	28	29	30
31	32	33	34	35	36	37	38	39	40
41	42	43	44	45	46	47	48	49	50
51	52	53	54	55	56	57	58	59	60

Skip count by 2s: 2, 4, 6, 8, 10, 12

Skip count by 5s: 5, 10, 15, 20, 25, 30

Skip count by 10s: 10, 20, 30, 40, 50, 60

Skip count. Write the missing numbers.

1. Skip count by 2s: 6, ______, 10, ______, 14, ______

2. Skip count by 5s: 15, 20, ______, 30, ______, 40

3. Skip count by 10s: ______, 50, 40, ______, 20, ______

Talk Math When would you use skip counting?

Name ....

CCSS

On My Own

Use the hundred chart to skip count.

1	2	3	4	5	6	7	8	9	10
11	12	13	14	15	16	17	18	19	20
21	22	23	24	25	26	27	28	29	30
31	32	33	34	35	36	37	38	39	40
41	42	43	44	45	46	47	48	49	50
51	52	53	54	55	56	57	58	59	60
61	62	63	64	65	66	67	68	69	70
71	72	73	74	75	76	77	78	79	80
81	82	83	84	85	86	87	88	89	90
91	92	93	94	95	96	97	98	99	100

4. Start on 5. Count by 5s. Color the numbers blue.

5. Start on 10. Count by 10s. Circle the numbers red.

Write the missing number. Describe the pattern.

6. 34, 36, 38, ______

Skip count by ______.

7. 10, ______, 30, 40

Skip count by ______.

8. 24, 22, ______, 18

Skip count by ______.

9. 40, 35, ______, 25

Skip count by ______.

Problem Solving

10. If there are 2 vases with 10 flowers in each vase, how many flowers are there in all?

______ flowers

11. There are 3 fish bowls. How many fish are there in all if there are 5 fish in each bowl?

______ goldfish

12. How many frogs are there in all if there are 3 lily pads and each one has 2 frogs?

______ frogs

Write an addition story for the cubes.

____

____

____

Name ....

Operations and Algebraic Thinking
2.OA.1, 2.OA.2, 2.NBT.2

CCSS

My Homework

Lesson 1
Skip Count on a Hundred Chart

Homework Helper

Need help? connectED.mcgraw-hill.com

You can skip count on a hundred chart.

Skip count by 2s.
58, 60, 62, 64, 66, 68

Skip count by 5s.
55, 60, 65, 70, 75, 80

Skip count by 10s.
60, 70, 80, 90, 100

51	52	53	54	55	56	57	58	59	60
61	62	63	64	65	66	67	68	69	70
71	72	73	74	75	76	77	78	79	80
81	82	83	84	85	86	87	88	89	90
91	92	93	94	95	96	97	98	99	100

Practice

Skip count. Write the missing numbers.

1. 62, _______, 66, _______, _______, 72

2. 60, _______, 70, _______, 80, _______

3. 95, _______, _______, 80, 75, _______

4. 88, _______, 84, _______, _______, 78

Write the missing number. Describe the pattern.

5. 50, _______, 70, 80

Skip count by _______.

6. 30, 35, _______, 45

Skip count by _______.

7. 75, _______, 85, 90

Skip count by _______.

8. 90, _______, _______, 60

Skip count by _______.

9. **Draw a picture to solve.**

There are 20 wheels on a group of bicycles. Each bike has 2 wheels. How many bikes are there?

_________ bicycles

Vocabulary Check

10. Which is not an example of a way to **skip count**? Circle it.

10, 12, 16, 20, 24   15, 20, 25, 30, 35

28, 26, 24, 22, 20   40, 50, 60, 70, 80

Math at Home Practice skip counting by 2s, 5s and 10s. Use cereal or beans.

Name ....

Skip Count by 2s, 5s, and 10s

Lesson 2

ESSENTIAL QUESTION
How can equal groups help me add?

Explore and Explain

This desert is not big enough for all of us.

____, ____, ____ snakes

Teacher Directions: Use counters to model. There are 3 groups of snakes in the desert. There are 5 snakes in each group. Skip count. Write the numbers. How many snakes in all?

See and Show

Equal groups have the same number of objects. Put two counters in each group. How many equal groups are there?

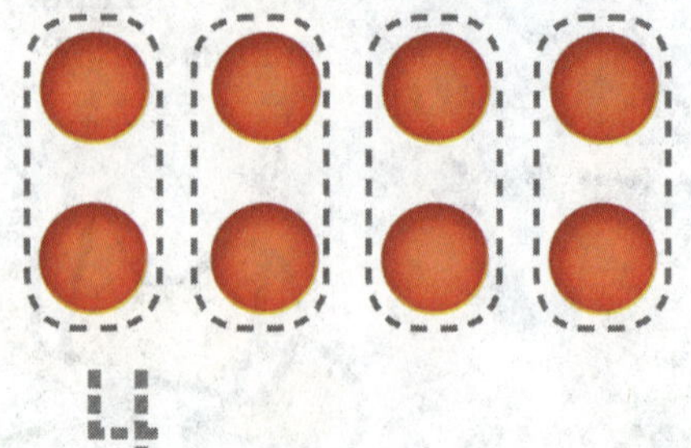

4 equal groups

Helpful Hint
When there are equal groups, you can skip count to find the total.

Skip count to find the total.

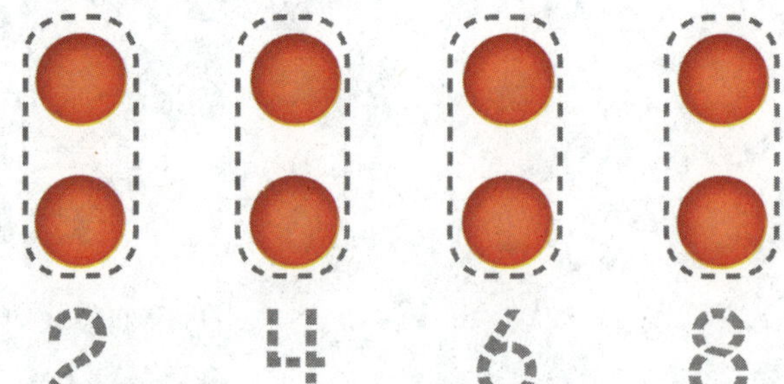

2, 4, 6, 8 counters in all

Use ● to model equal groups.
Skip count to find the total.

1. 3 groups of 2 ____ in all
2. 4 groups of 5 ____ in all

Describe the groups. Skip count to find the total.

3. 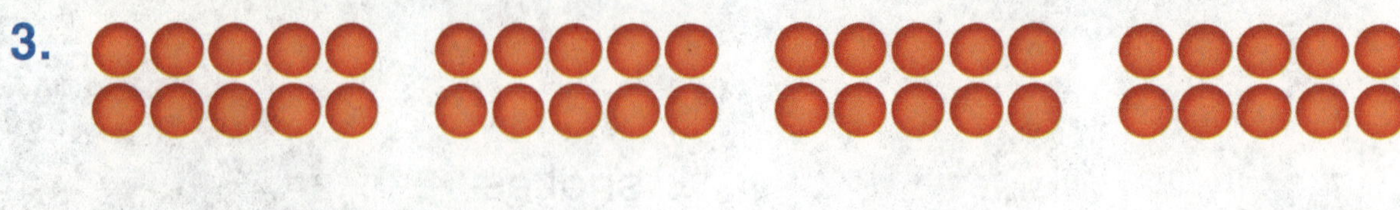

____ groups of ____ ____ in all

Talk Math Create a skip-counting story for Exercise 3.

CCSS

Name ______________________________

On My Own

Use ● to model equal groups.
Skip count to find the total.

4. 2 groups of 10 _____ in all

5. 6 groups of 2 _____ in all

Describe the groups. Skip count to find the total.

6.

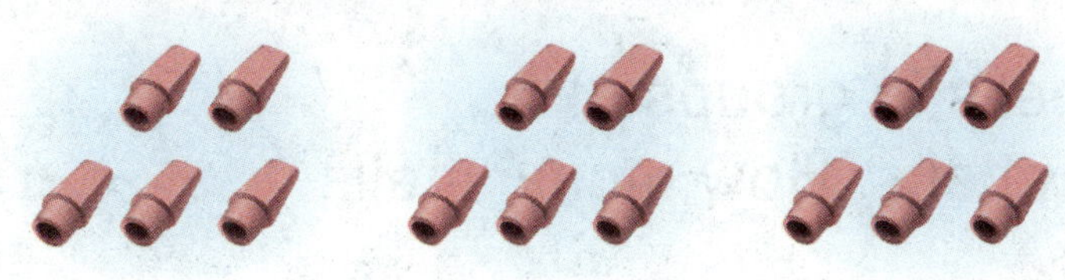

_______ groups of _______ _______ in all

7.

_______ groups of _______ _______ in all

8.

_______ groups of _______ _______ in all

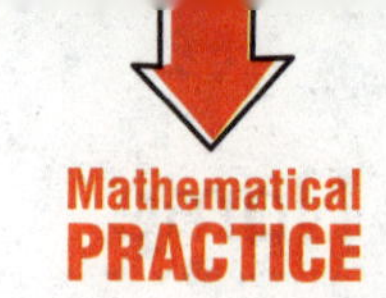

Problem Solving

Solve. Draw a picture to help, if needed.

9. Lori has 6 bunches of grapes. Each bunch has 10 grapes. How many grapes does Lori have in all?

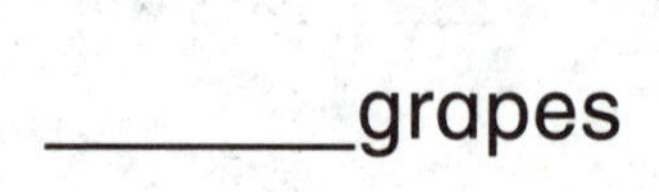

_______ grapes

10. Paul organized his shells in groups of 5. He has 3 groups of 5 shells. How many shells does he have in all?

_______ shells

11. Myla put 10 cookies each in 4 bags for a bake sale. Kate put 10 cookies in each of 3 bags. How many cookies do they have in all?

_______ cookies

How can you find the total number of tennis balls in 4 groups of 10 tennis balls? Use your vocabulary word.

____

____

____

Name ....

My Homework

Lesson 2
Skip Count by 2s, 5s and 10s

Homework Helper

eHelp Need help? connectED.mcgraw-hill.com

Make equal groups and skip count to find the total.

3 equal groups

2 4 6 owls in all

Practice

Skip count to find the total.

1.

_______ groups of _______ _______ spiders in all

2.

_______ groups of _______ _______ turtles in all

Solve. Draw a picture to help, if needed.

3. Evan has 6 groups of 5 toy cars. How many toy cars does he have in all?

_______ toy cars

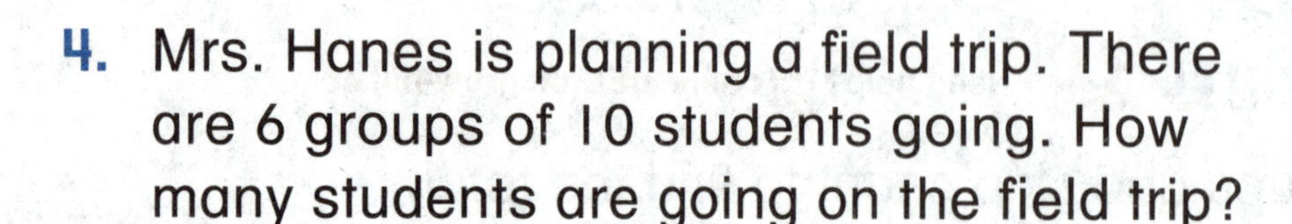

4. Mrs. Hanes is planning a field trip. There are 6 groups of 10 students going. How many students are going on the field trip?

_______ students

5. Austen has 6 sets of 10 baseball cards. Tim has 3 sets of 10 baseball cards. How many baseball cards do they have in all?

_______ baseball cards

Vocabulary Check

6. Circle **equal groups** of five buttons.

Math at Home Use buttons, macaroni, or pennies. Have your child make equal groups of 2s, 5s, and 10s. Have him or her skip count to find the totals.

Name

CCSS

Problem Solving

STRATEGY: Find a Pattern

Lesson 3

ESSENTIAL QUESTION

How can equal groups help me add?

Kelly puts 5 party favors in each bag. She has 6 bags. How many favors does she need in all?

1 Understand

Underline what you know.
Circle what you need to find.

2 Plan

How will I solve the problem?

3 Solve

Find a pattern.

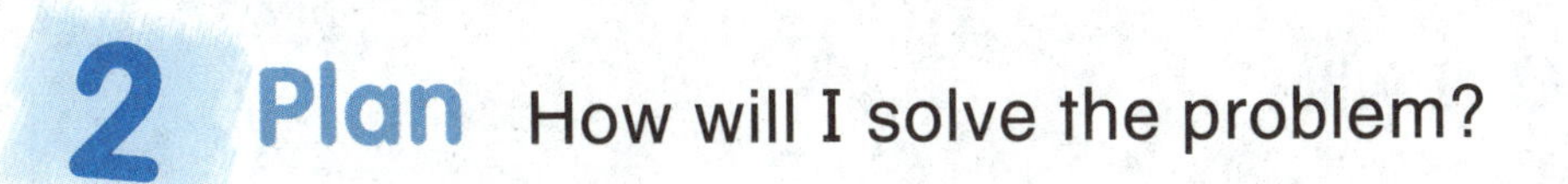

5, 10, 15, 20, 25, 30
bag 1, bag 2, bag 3, bag 4, bag 5, bag 6

She needs 30 favors in all.

4 Check

Is my answer reasonable? Explain.

Practice the Strategy

Kate and Pedro count the tickets they sold to attend a museum. They sold 10 tickets each day for 5 days. How many tickets were sold?

1 Understand Underline what you know.
Circle what you need to find.

2 Plan How will I solve the problem?

3 Solve I will . . .

_____, _____, _____, _____, _____

By day 5, _____ tickets were sold in all.

4 Check Is my answer reasonable? Explain.

Name ....

Mathematical PRACTICE

CCSS

Apply the Strategy

1. Xavier is looking at a map of the desert. He knows that each finger width is about 10 miles. How many miles will he count if he uses 7 finger widths?

_______ miles

2. Sam thinks of the number pattern 15, 20, 25, 30. He continues this pattern. What will be the next four numbers?

_______, _______, _______, _______

3. Amar sees this number pattern. What is the missing number?

10, _______, 20, 25, 30

Review the Strategies

Choose a strategy

- Write a number sentence.
- Act it out.
- Find a pattern.

4. Justin is stacking books. There are 5 books in the first stack. 10 in the second stack. The third stack has 15 books. The pattern continues. How many books will be in the next stack?

_____ books

5. On the first day of the food drive, Ms. Buckle's class collects 8 cans. Mr. Cline's class collects 6 cans, and Mrs. Brown's class collects 5 cans. How many cans do they collect in all?

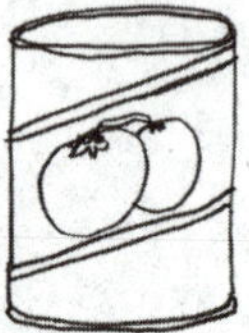

_____ cans

6. Josh recorded how many inches of snow fell in each month. Continue the pattern. How many inches fell in January?

Month	Inches
November	14
December	16
January	
February	20

_____ inches

Name ....

My Homework

Lesson 3
Problem-Solving: Find a Pattern

Deb paints 2 flowers in the first row, 4 flowers in the second row, and 6 flowers in the third row. If she continues the pattern, how many flowers will be in the fifth row?

1 Understand

Underline what you know.
Circle what you need to find.

2 Plan

How will I solve the problem?

It's a masterpiece!

3 Solve

I will find a pattern.

2,	4,	6,	8,	10
Row 1,	Row 2,	Row 3,	Row 4,	Row 5

There are 10 flowers in the fifth row.

4 Check

Is my answer reasonable?

Practice

Underline what you know. Circle what you need to find. Find a pattern to solve.

1. One elephant has four legs.
   Two elephants have eight legs.
   How many legs do five elephants have? _______

2. Corey is thinking of a number.
   It is the next number in the pattern
   50, 45, 40, 35. What is the number? _______

3. Brad has groups of 10 pennies.
   Brad counts 40, but he has 6 more stacks to count. How many pennies does he have?

   _______ pennies

Test Practice

4. Which of the following is an example of skip counting?

8, 10, 20, 30, 40 ○	40, 35, 25, 15, 20 ○
20, 25, 30, 35, 40 ○	45, 55, 60, 70, 80 ○

Math at Home Ask your child to write a number pattern that uses 5 or 6 numbers. Have him or her explain the pattern to you.

Name ______________________________

Check My Progress

Vocabulary Check

Complete each sentence.

skip count **equal groups**

1. When you have sets of the same number of objects you have ______________.

2. You can ______________ to help you count equal groups.

Concept Check

Skip count. Write the missing numbers.

3. 6, ____, ____, 12, 14, ____, 18, ____

4. 15, 20, ____, 30, ____, 40, ____, 50

5. 20, ____, ____, 50, ____, ____, 80, 90

6. ____, 55, ____, 65, ____, ____, 80

7. 15, 20, ____, 30

Skip count by ____.

8. 20, ____, 16, 14

Skip count by ____.

Describe the groups. Skip count to find the total.

9.

____ groups of ____ ____ ants in all

10.

____ groups of ____ ____ lizards in all

11. Jason has 3 groups of 5 baseball cards. Cameron has 2 groups of 10 baseball cards. Which boy has more cards?

____

Test Practice

12. Jose has 3 boxes of rocks in his collection. Each box holds 10 rocks. He fills one more box. How many rocks does he have in all?

10	13	30	40
○	○	○	○

Name

Repeated Addition

Lesson 4

ESSENTIAL QUESTION
How can equal groups help me add?

Explore and Explain

_____ + _____ + _____ + _____ = _____

Teacher Directions: Place a connecting cube on each blade of the wind turbines. Write how many cubes are on each turbine. Count all the connecting cubes. How many blades in all? Write the total for the number sentence.

See and Show

When groups are equal, you can use **repeated addition** to find the total.

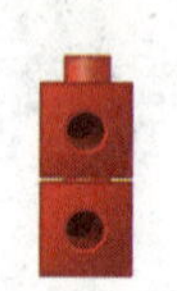
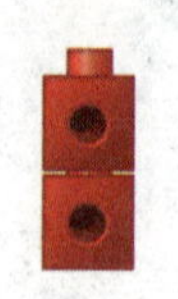
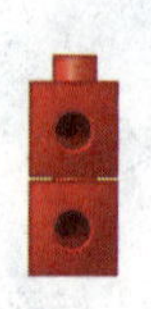

Helpful Hint
You can skip count to find sums like this.

2 + 2 + 2 + 2 = 8

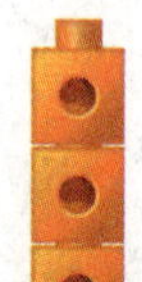
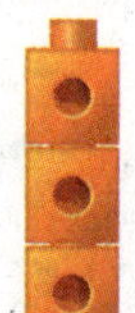

3 + 3 + 3 = 9

Use connecting cubes to model equal groups. Add.

1. 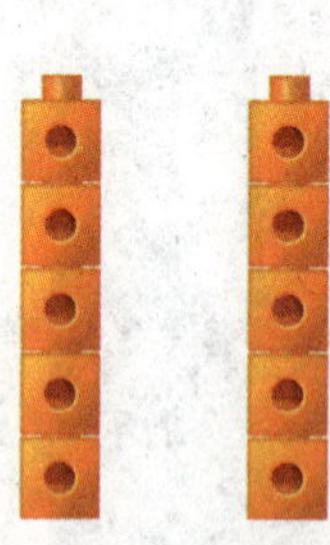

___ + ___ = ___

2. 

___ + ___ + ___ = ___

3. ___ + ___ = ___

Talk Math Create a story for 2 + 2 + 2 + 2 + 2.

Name ______________________

CCSS

On My Own

Add.

4.

___ + ___ + ___ = ___

5.

___ + ___ + ___ + ___ + ___ = ___

6.

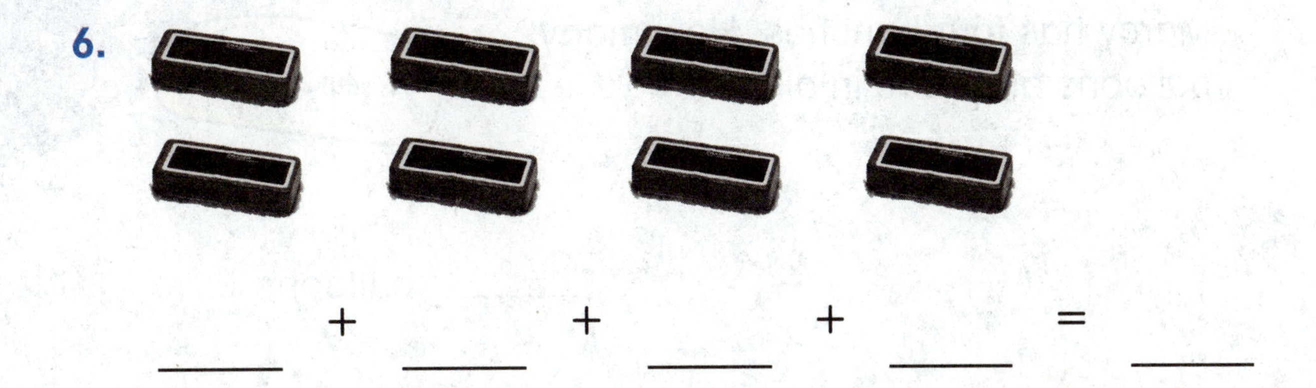

___ + ___ + ___ + ___ = ___

7. Draw your own example. Then add.

4 + 4 + 4 + 4 = ___

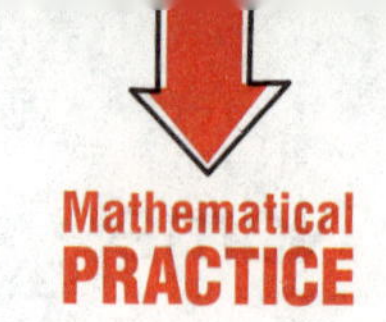

Problem Solving

Use repeated addition to solve.

8. Mike has five pairs of socks. Each pair has two socks. How many socks does Mike have?

_______ socks

9. Brad makes four groups of animal cards. Each group has three cards. How many cards does he make?

_______ cards

10. There are four balloons in each bunch. Marcy has four bunches. How many balloons are there in all?

Oh, no! a cactus!

_______ balloons

HOT Problem Jaya writes a repeated number sentence. It has three numbers. The sum is 15. What is the number sentence? Explain.

____

____

____

Name ______________________

My Homework

Lesson 4
Repeated Addition

Homework Helper

Need help? connectED.mcgraw-hill.com

When groups are equal, use repeated addition to find the total.

2 + 2 + 2 = 6 cactus

Practice

Add.

1\.

_____ + _____ + _____ + _____ = _____

2\.

_____ + _____ + _____ = _____

Add.

3.

_____ + _____ = _____ lady bugs

Use repeated addition to solve.

4. Marco has 4 fish tanks. Each tank has 10 fish. How many fish does Marco have in all?

_____ fish

5. Mandy has 4 boxes of raisins. Each box has 5 raisins. How many raisins does Mandy have?

_____ raisins

Vocabulary Check

6. Circle the **repeated addition** sentence.

5, 10, 15, 20

9 + 3 + 9

3 + 3 + 3

Math at Home Have your child put beans or pennies in 5 equal groups of 6 to show repeated addition. Ask them to write a number sentence and solve.

Name ....

Repeated Addition with Arrays

Lesson 5

ESSENTIAL QUESTION
How can equal groups help me add?

Explore and Explain

Jamar					
Meko					
Kyle					
Laurel					

Teacher Directions: Jamar and 3 of his friends each collect 2 rocks. How many rocks do they have in all? Use connecting cubes to show equal groups of rocks. Color in the grid to show your work. Write a number sentence to show the total.

See and Show

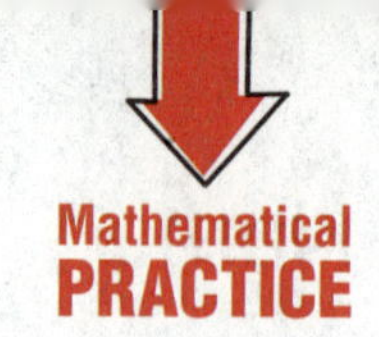

In an **array**, objects are shown in rows and columns.

3 rows of 4

3 rows of 4

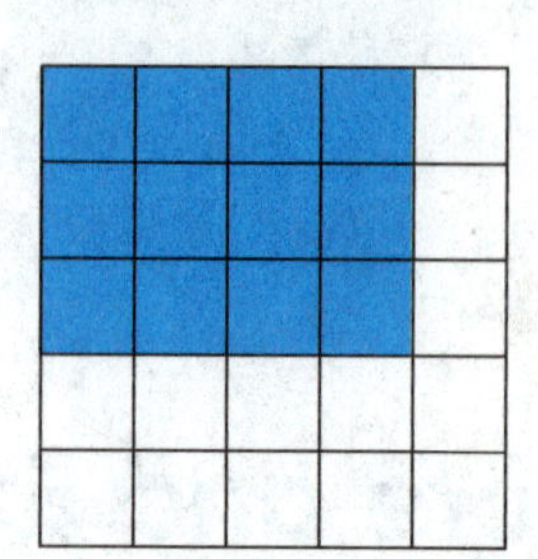

4 + 4 + 4 = 12

Describe each array using a number sentence.

1. 

___ + ___ + ___ = ___

___ rows of ___ turtles

2. 

___ + ___ = ___

___ rows of ___ cacti

3. Shade the grid to show 4 rows of 2.
Write a number sentence to describe it.

____

Talk Math How can arrays help you add?

Name

On My Own

Describe the array using a number sentence.

4. 

____ + ____ + ____ = ____ ____ rows of ____ lizards

Shade each grid to show the array. Write a number sentence to describe it.

5. Show 4 rows of 3.

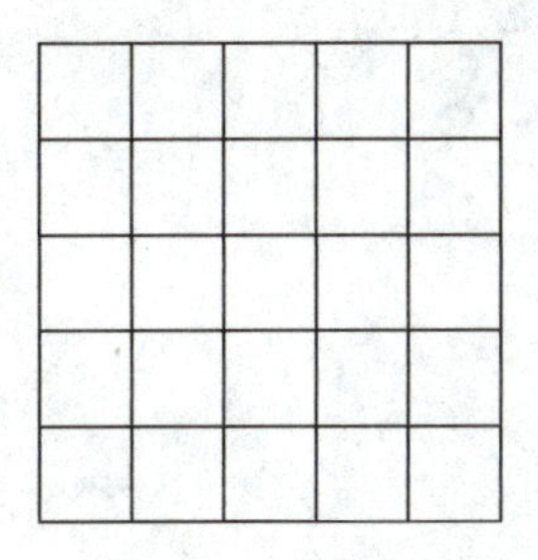

____

6. Show 4 rows of 1.

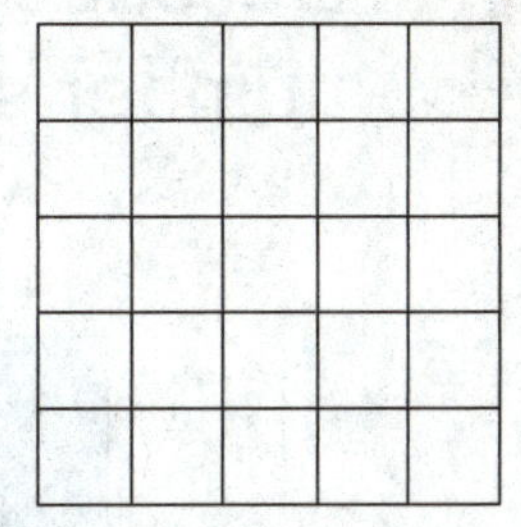

____

7. Show 2 rows of 5.

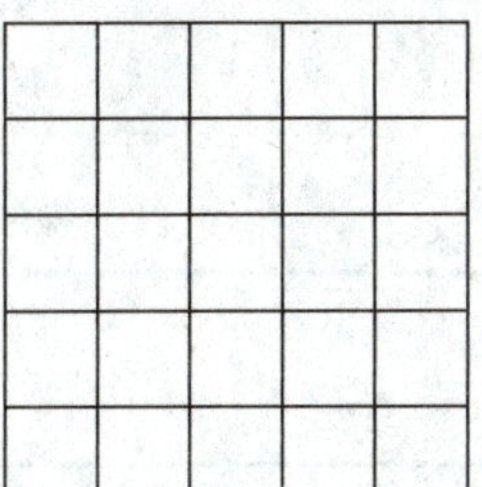

____

8. Show 3 rows of 4.

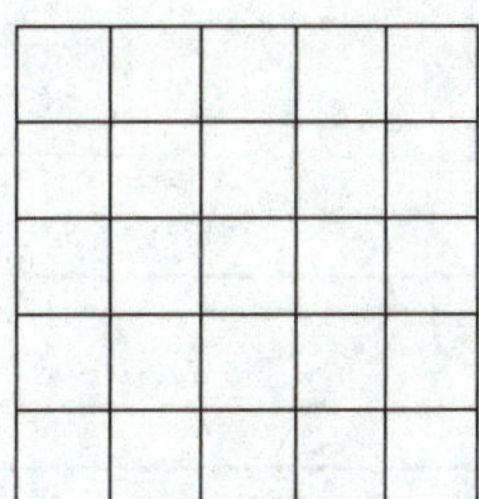

____

Problem Solving

Mathematical PRACTICE

Use an array to solve.

9. Kathy puts 5 chairs in one row. She puts the same number of chairs in three more rows. How many chairs are there in all?

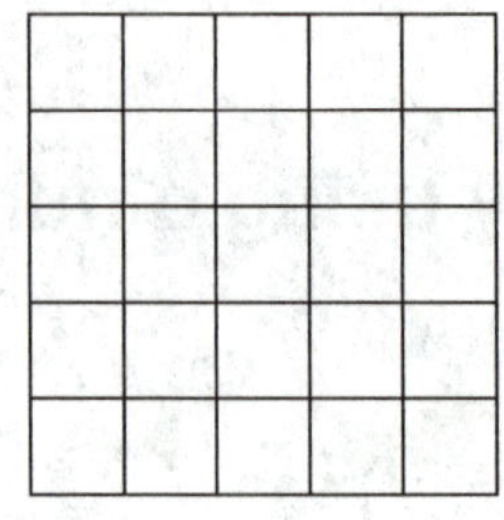

I call dibs on this one.

_____ chairs

10. Three hills each have 4 wind turbines. How many wind turbines are there in all?

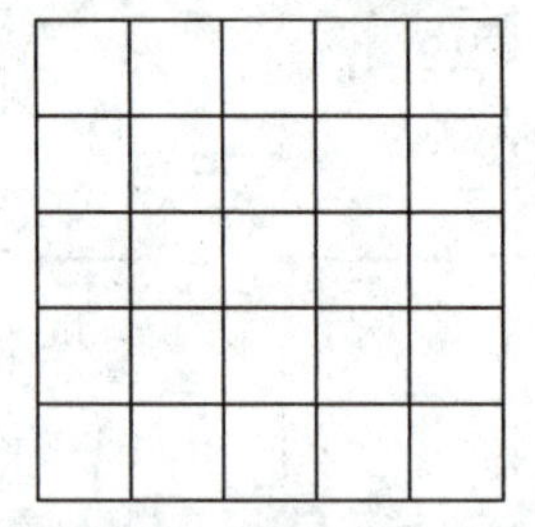

_____ wind turbines

Describe how this array shows the number sentence $2 + 2 + 2 = 6$.

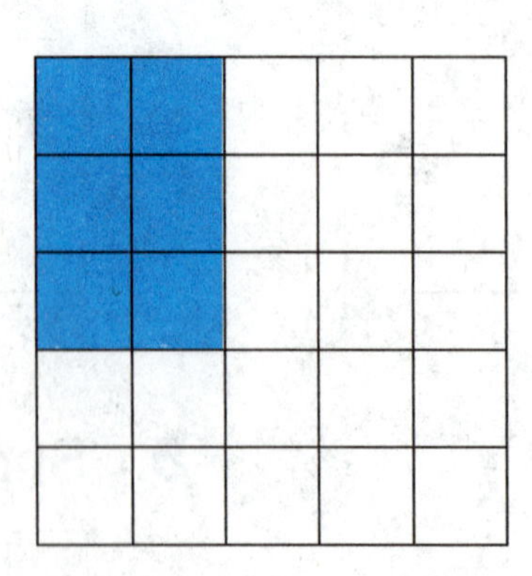

____

____

____

Name

My Homework

Lesson 5
Repeated Addition with Arrays

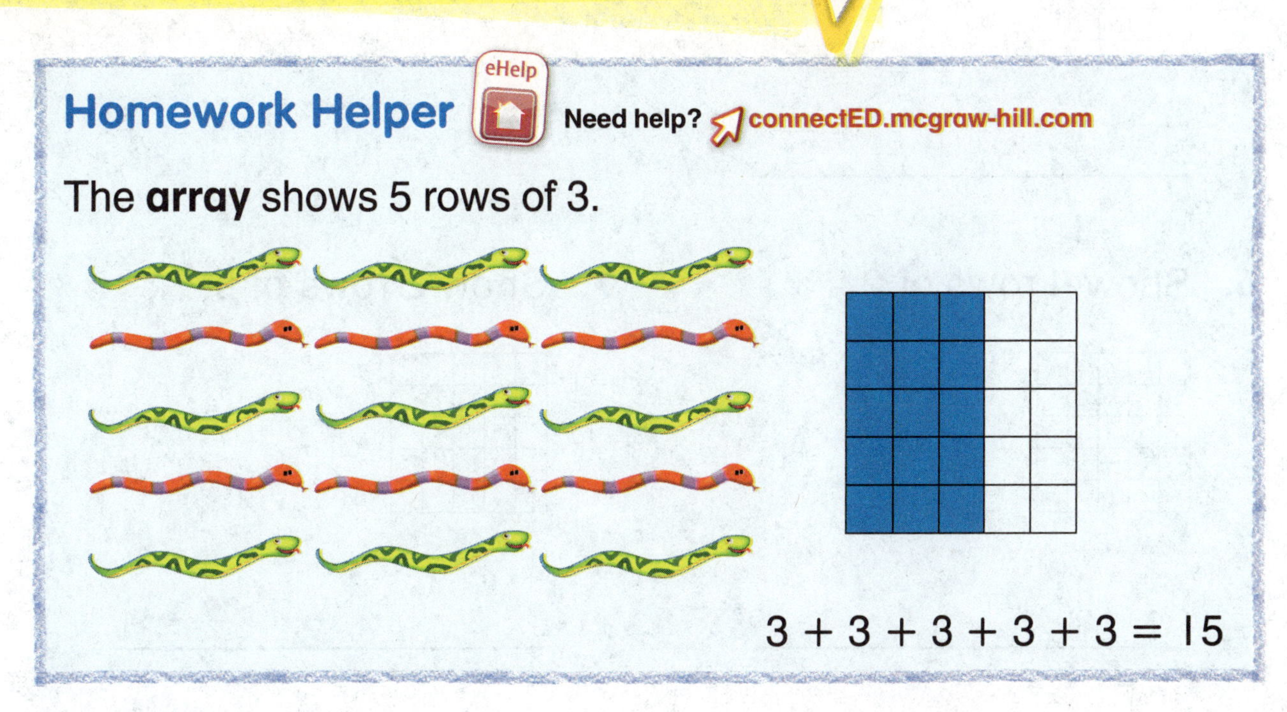

Homework Helper

eHelp Need help? connectED.mcgraw-hill.com

The **array** shows 5 rows of 3.

3 + 3 + 3 + 3 + 3 = 15

Practice

Describe each array using a number sentence.

1.    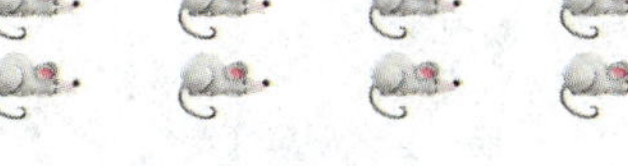

___ + ___ + ___ = ___

___ rows of ___ mice

2.     

___ + ___ = ___

___ rows of ___ owls

3. Show 3 rows of 5.

____

4. Show 3 rows of 2.

____

Describe each array using a number sentence.

5. Show 3 rows of 5.

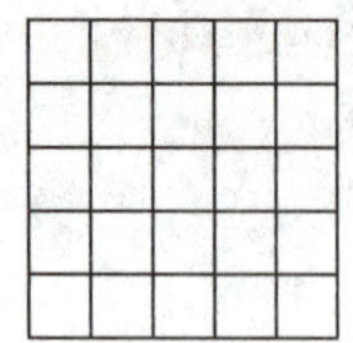

____

6. Show 4 rows of 2.

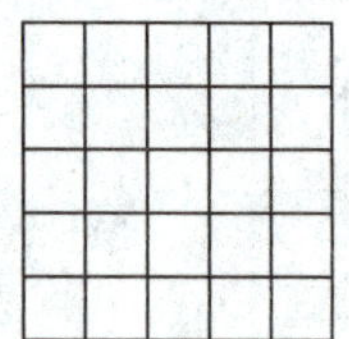

____

7. Show 2 rows of 4.

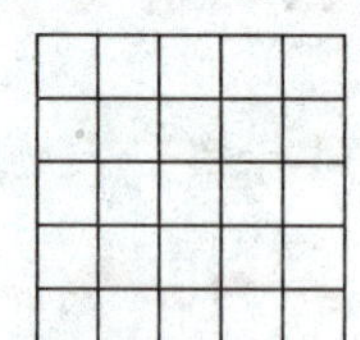

____

8. There are 4 legs on 1 camel. How many legs are on 5 camels?

_______ legs in all

Vocabulary Check

9. Draw an **array** of 2 rows of 3.

Math at Home Have your child think of things around them that are in an array. Ex: eggs in a carton, days on the calendar, etc.

Name

Even and Odd Numbers

Lesson 6

ESSENTIAL QUESTION
How can equal groups help me add?

Remember, leftovers mean it's odd!

Explore and Explain

Try to break the cubes into two equal groups.

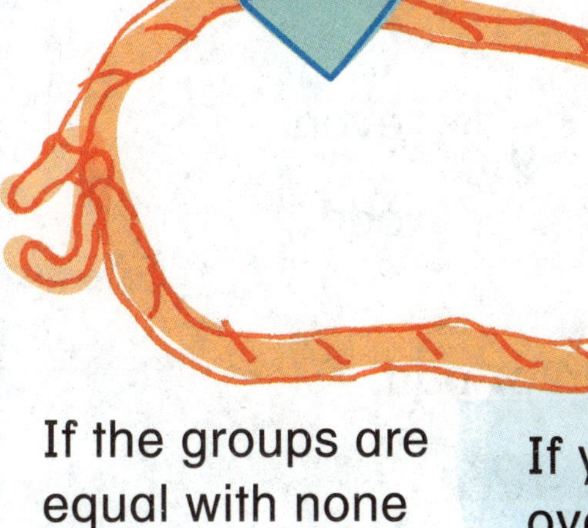

If the groups are equal with none left over, your number is even.

If you have one left over cube, your number is odd.

Teacher Directions: Roll a blue number cube. Make a cube train to show the number. Try to break the train into two equal groups. Place any extras on the red cube. Is the number even or odd? Repeat. Draw the groups.

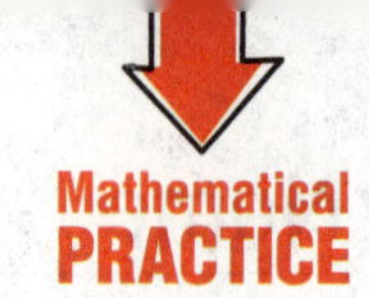

Mathematical PRACTICE

See and Show

An **even** number of objects can be counted by 2s.

4 2 4

Helpful Hint
Start with 4 cubes.
Make pairs. Count. 2, 4

An **odd** number of objects has 1 left over when counted by 2s.

5 2 4 5

Start with 5 cubes.
Make pairs. 1 is left over. Count. 2, 4, 5

Use connecting cubes to show the number. Circle *even* or *odd*.

1. 4	even odd	**2.** 5	even odd
3. 8	even odd	**4.** 9	even odd

Write the number of cubes. Circle even or odd.

5. ______ even odd

6. ______ even odd

When might you use even and odd numbers?

Name ____________________

On My Own

Use connecting cubes to show the number. Circle *even* or *odd*.

7. 11 even odd

8. 12 even odd

9. 17 even odd

10. 14 even odd

11. 8 even odd

12. 19 even odd

Write the number of cubes. Circle *even* or *odd*.

13.

______ even odd

14.

______ even odd

15.

______ even odd

16.

______ even odd

17. 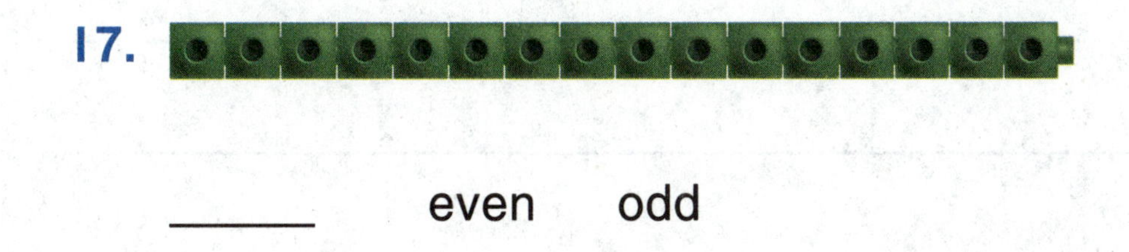

______ even odd

Problem Solving

18. Ben has 16 pennies. Does he have an even or odd number of pennies?

____

19. Mr. Rice thinks of a number between 14 and 17. The number is even. What is the number?

______

20. Miss Lee has 18 plants. She wants to split them between two friends. How many plants will she give each friend?

______ plants

Write Math Explain how you know 15 is an odd number.

____

____

____

Name ....

My Homework

Lesson 6
Even and Odd Numbers

Homework Helper

eHelp Need help? connectED.mcgraw-hill.com

An even number of objects can be counted by 2s.

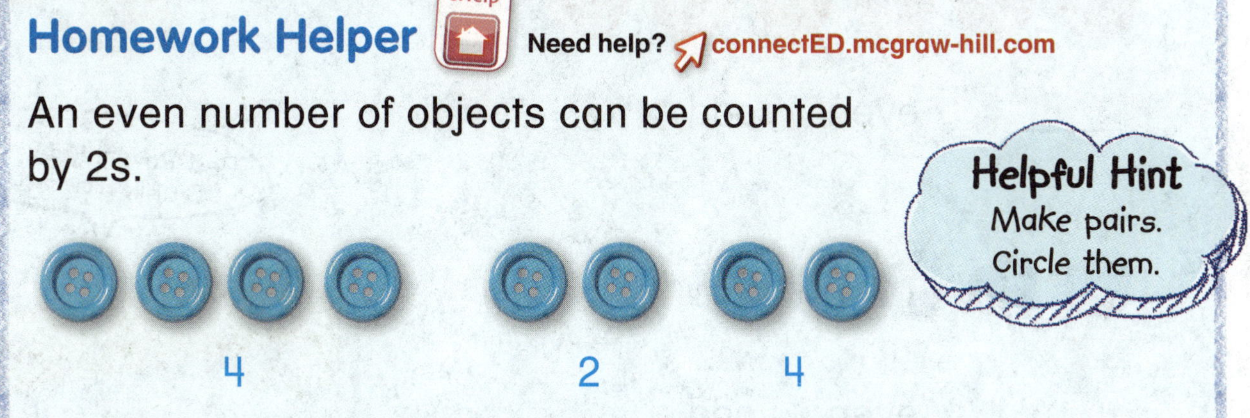

An odd number of objects has 1 left over when counted by 2s.

Practice

Use buttons or pennies to show each number. Circle *even* or *odd*.

1. 8 even odd

2. 16 even odd

3. 11 even odd

4. 19 even odd

Write the number of buttons. Circle *even* or *odd*.

> **Helpful Hint**
> Circle pairs. If none are left over, the number is even.

5.       

_____ even odd

6.     

_____ even odd

7.     

_____ even odd

8. Katy has 13 oranges, 12 apples, and 6 bananas. She also has two boxes. Which fruit cannot be put evenly into the two boxes?

_____

Vocabulary Check

Draw lines to match.

9. **odd** — A number that can be counted by 2s.

10. **even** — A number that can be counted by 2s and has one left over.

Math at Home Give your child random numbers up to 20 and let them tell you if they are even or odd. Use buttons or coins if needed.

Name

Sums of Equal Numbers

Lesson 7

ESSENTIAL QUESTION
How can equal groups help me add?

Explore and Explain

Tools

Teacher Directions: Place a train of 6 connecting cubes in the top box. Trace it. Break it into two equal groups. Put one group in each of the two bottom boxes. Trace them. Trace the number sentence.

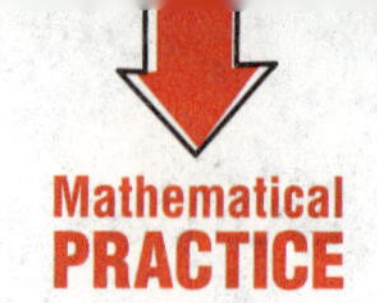

See and Show

Mathematical PRACTICE

Any even number can be written as the sum of two equal addends.

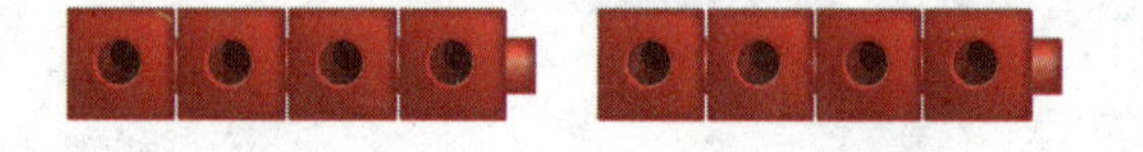

8 = 4 + 4

2 = 1 + 1

Write the equal addends that make each sum.

1. 6 = ______ + ______
2. 4 = ______ + ______
3. ______ + ______ = 2
4. 8 = ______ + ______
5. 14 = ______ + ______
6. ______ + ______ = 18

Find the missing number.

7. ______ + 4 = 8
8. 9 + ______ = 18

Talk Math How did you find the equal addends that make each sum?

Name ______________________________

On My Own

Write the equal addends that make each sum.

9. 2 = ____ + ____

10. 12 = ____ + ____

11. 10 = ____ + ____

12. 14 = ____ + ____

13. 18 = ____ + ____

14. 8 = ____ + ____

15. ____ + ____ = 6

16. ____ + ____ = 4

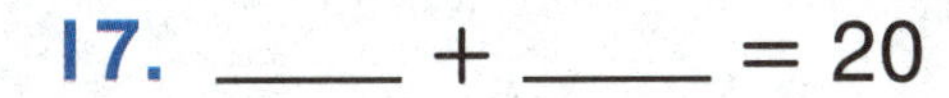

17. ____ + ____ = 20

18. ____ + ____ = 16

Find the missing addends.

19. 10 + ______ = 20

20. ______ + 6 = 12

21. 4 + ______ = 8

22. ______ + 3 = 6

23. ______ + 7 = 14

Problem Solving

24. Henry buys 6 apples. His mother buys the same amount. Write the number sentence. How many apples do they have?

_____ + _____ = _____

25. Luke has 10 puppies. An equal number of them are black and brown puppies. How many are brown?

_____ brown puppies

26. Janelle brought 7 magazines. The class now has 14 magazines. How many magazines did they already have?

_____ magazines

Write Math Tamika and Jen have the same number of beads. They have 18 beads in all. How many beads does each girl have? Explain.

____

____

____

Name ____________________

My Homework

Lesson 7
Sums of Equal Numbers

Homework Helper

Need help? connectED.mcgraw-hill.com

Any even number can be written as the sum of two equal addends.

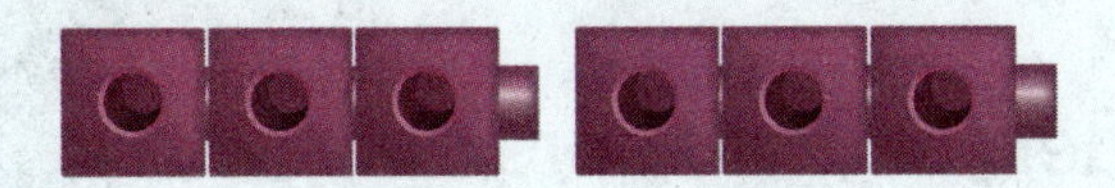

6 = 3 + 3

4 = 2 + 2

Practice

Write the equal addends that make each sum.

1. 10 = ______ + ______
2. 2 = ______ + ______
3. 18 = ______ + ______
4. 8 = ______ + ______

Find the missing number.

5. 7 + ______ = 14
6. ______ + 3 = 6
7. 8 + ______ = 16
8. ______ + 6 = 12

Find the missing addends.

9. 5 + ______ = 10

10. ______ + 2 = 4

11. Maria and Steve each have 8 toys for the toy drive. How many toys do they have in all?

______ toys

12. Mrs. Miller's class has 18 students. There are an equal number of boys and girls in the class. Write a number sentence to show how many boys and how many girls are in the class.

______ = ______ + ______

Test Practice

13. Kylie buys 7 bananas. Her mom also buys 7 bananas. How many bananas do they buy in all?

0	7	10	14
○	○	○	○

Math at Home Work with your child to create a number line from 1 to 20. Have your child circle 18, 14, and 10. Ask your child to tell you the equal addends for each of these numbers.

Name ____________________

My Review

Chapter 2

Number Patterns

Vocabulary Check

array **equal groups** **even**
odd **repeated addition** **skip count**

Complete each sentence.

1. ____________________ have the same number of objects.

2. You can ____________________ by counting objects in equal groups of two or more.

3. An ____________________ number of objects has 1 left over when counted by 2s.

4. An ____________________ is a group of objects arranged in rows and columns of the same length.

Concept Check

Write the missing numbers.

5. 65, ______, 75, 80, ______, 90, ______, 100

6. 30, ______, 50, 60, ______, 80, ______

Fill in the blanks to describe each array.

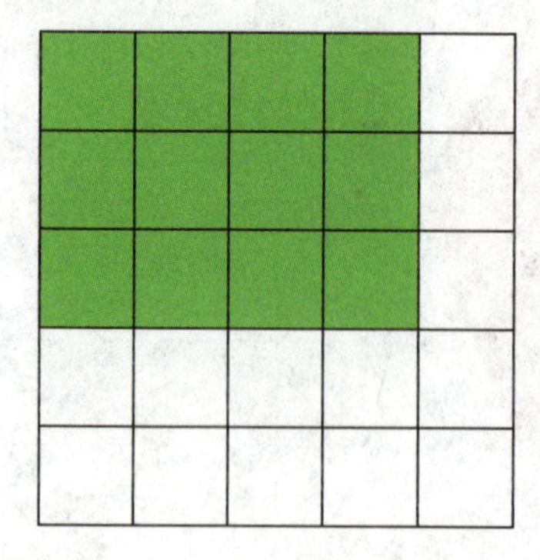

______ rows of ______

8.

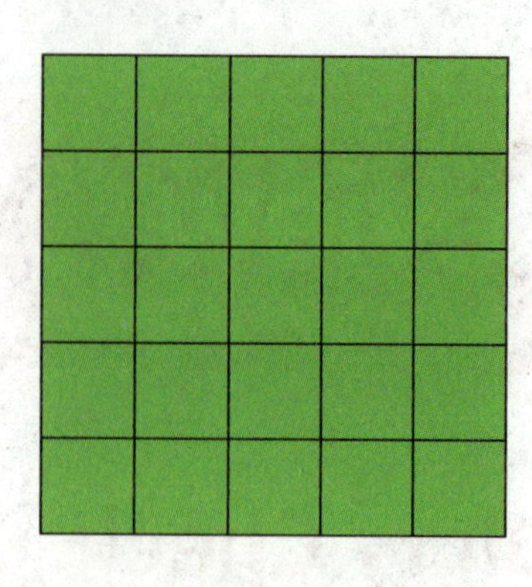

______ rows of ______

Circle *even* or *odd*.

9. 8 even odd

10. 13 even odd

11. 10 even odd

Write the equal addends that make each sum.

12. 8 = ______ + ______

13. 16 = ______ + ______

14. 12 = ______ + ______

15. 18 = ______ + ______

Find the missing addends.

16. 7 + ______ = 14

17. ______ + 9 = 18

18. ______ + 6 = 12

19. 3 + ______ = 6

20. ______ + 5 = 10

21. 7 + ______ = 14

Name ....

Problem Solving

22. Liam has 5 dogs. Each dog has 4 bones. How many bones do the dogs have in all?

_____ bones

23. Lauren has 3 pages of stamps. Each page has 5 stamps on it. How many stamps does Lauren have in all?

_____ stamps

24. Jonah has an even number of golf balls. He has more than 13 and less than 16 golf balls. How many golf balls does he have?

_____ golf balls

Test Practice

25. Katherine has skip counted 16 rubber bracelets by 2s. If she continues skip counting by 2s, how many does she have after counting 5 more sets?

20	21	24	26
◯	◯	◯	◯

Reflect

Chapter 2
Answering the Essential Question

Show how equal groups help you.

Write a repeated addition sentence that describes the cubes.

____

Shade the array to show 4 rows of 5.

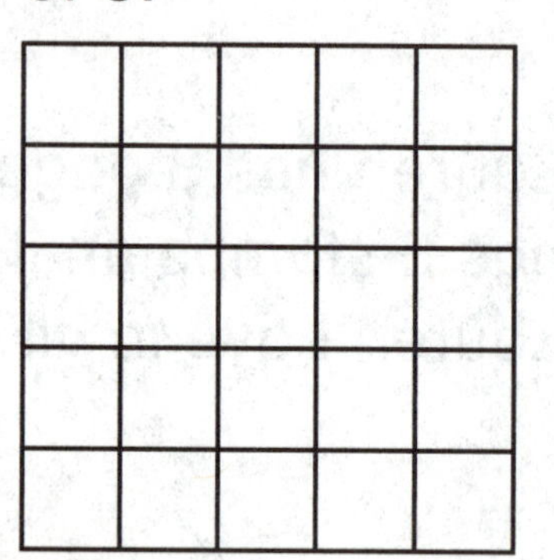

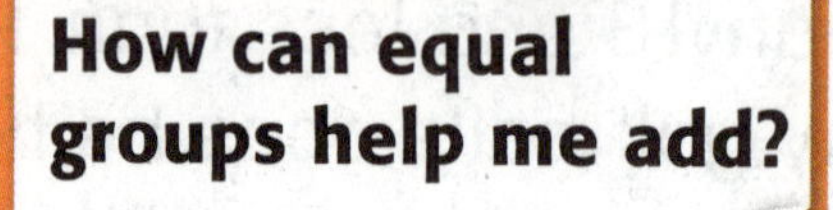

Skip count by 5s. What are the missing numbers?

40, ________, ________, 55

Circle pairs.

Is the number of counters even or odd?

____

Chapter

3 Add Two-Digit Numbers

ESSENTIAL QUESTION

How can I add two-digit numbers?

I Like Teamwork!

Watch a video!

Watch

My Common Core State Standards

Operations and Algebraic Thinking

2.OA.1 Use addition and subtraction within 100 to solve one-and two-step word problems involving situations of adding to, taking from, putting together, taking apart, and comparing, with unknowns in all positions

Number and Operations in Base Ten *This chapter also addresses these standards:*

2.NBT.5 Fluently add and subtract within 100 using strategies based on place value, properties of operations, and/or the relationship between addition and subtraction.

2.NBT.6 Add up to four two-digit numbers using strategies based on place value and properties of operations.

2.NBT.9 Explain why addition and subtraction strategies work, using place value and the properties of operations.

Standards for Mathematical PRACTICE

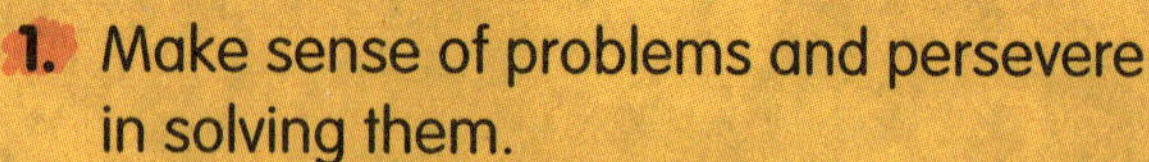

1. Make sense of problems and persevere in solving them.
2. Reason abstractly and quantitatively.
3. Construct viable arguments and critique the reasoning of others.
4. Model with mathematics.
5. Use appropriate tools strategically.
6. Attend to precision.
7. Look for and make use of structure.
8. Look for and express regularity in repeated reasoning.

= focused on in this chapter

Name

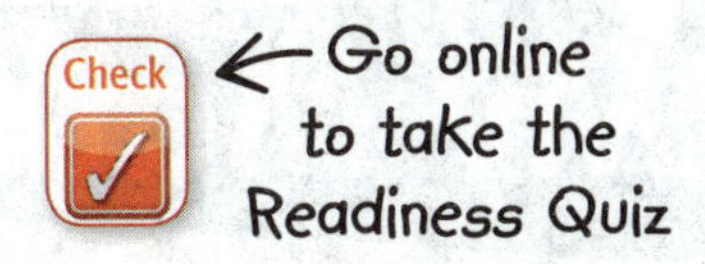

Write each number two ways.

1. 18 ones = ______

______ ten ______ ones

2. 26 ones = ______

______ tens ______ ones

Add.

3. $2 + 7$

4. $3 + 4$

5. $2 + 2$

6. $9 + 6$

7. $3 + 3 + 1 =$ ______

8. $4 + 2 + 3 =$ ______

9. Jack has 5 red marbles. Trent has 3 blue marbles. Lien has 2 green marbles. How many marbles do they have altogether?

______ marbles

Shade the boxes to show the problems you answered correctly.

1	2	3	4	5	6	7	8	9

Name

My Math Words

Review Vocabulary

addends   sum   add

Use the review words to complete the graphic organizer.

$5 + 6 = 11$

The blue number is called a ________.

The red numbers are called ________.

To join together to find the total is to ________.

Write a word problem about the addition sentence above.

____

____

____

My Vocabulary Cards

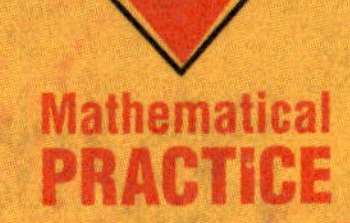

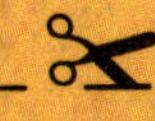

Lesson 3-2

regroup

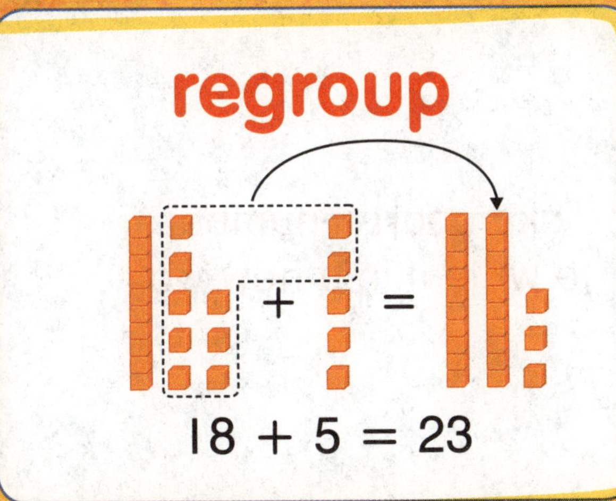

18 + 5 = 23

Directions:
Ideas for Use

- Have students write a tally mark on the card each time they read the word in this chapter.
- Ask students to use the blank cards to write their own vocabulary cards.

Take apart a number to write it in a new way.

My Foldable

FOLDABLES® Follow the steps on the back to make your Foldable.

$15 + 28 =$

$$\begin{array}{r} 22 \\ +\ 36 \\ \hline \end{array}$$

$54 + 37 =$

$$\begin{array}{r} 45 \\ +\ 19 \\ \hline \end{array}$$

$82 + 16 =$

$$\begin{array}{r} 31 \\ +\ 47 \\ \hline \end{array}$$

tens	ones

FOLDABLES®
Study Organizer

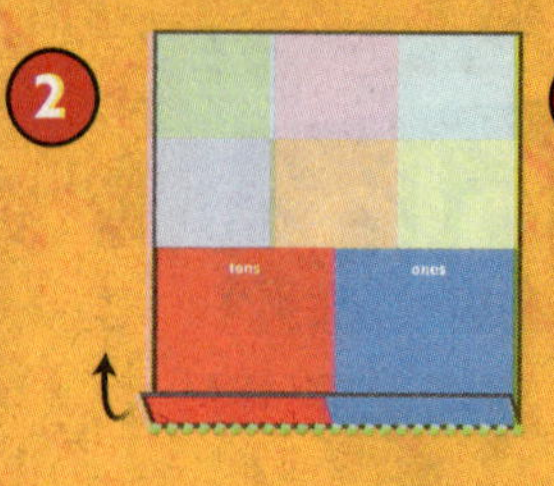

Name

Take Apart Tens to Add

Lesson 1

ESSENTIAL QUESTION
How can I add two-digit numbers?

Explore and Explain

We'll win if we pull together!

_____ + _____ = _____

Teacher Directions: Use base-ten blocks. Show 48 in the red box. Show 16 in the yellow box. Move some base-ten blocks from the yellow box to the red box to make 50. Write the number shown in each box. Write the number sentence.

See and Show

Numbers that end in zero are easier to add.
Take apart addends to make numbers that end in zero.

Find 28 + 36.

One Way:

Take apart 36 as 2 + 34.
Then make 28 + 2 into 30.

28 + 36

28 + 2 + 34

30 + 34 = 64

Another Way:

Take apart 28 as 24 + 4.
Then make 4 + 36 into 40.

28 + 36

24 + 4 + 36

24 + 40 = 64

So, 28 + 36 = 64.

Take apart an addend to solve.

1. 18 + 35

18 + 2 + 33

20 + 33 = 53

So, 18 + 35 = 53.

2. 44 + 26

44 + ____ + ____

____ + ____ = ____

So, 44 + 26 = ____.

How do you decide which addend to take apart when adding?

Name ____________________

I wonder what this take apart will be?

On My Own

Take apart an addend to solve.

3. 36 + 45

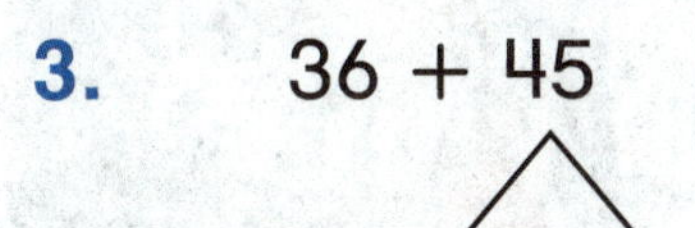

36 + _____ + _____

_____ + _____ = _____

So, 36 + 45 = _____.

4. 25 + 58

_____ + _____ + 58

_____ + _____ = _____

So, 25 + 58 = _____.

5. 67 + 26

67 + _____ + _____

_____ + _____ = _____

So, 67 + 26 = _____.

6. 38 + 14

38 + _____ + _____

_____ + _____ = _____

So, 38 + 14 = _____.

Problem Solving

Mathematical PRACTICE

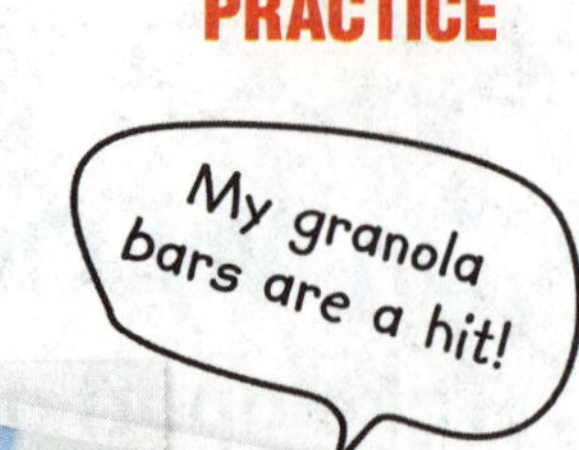

7. 28 people came to the bake sale in the morning. 34 people came to the bake sale in the afternoon. How many people came to the bake sale?

______ people

8. Bill collected 16 toys for the toy drive. Lisa collected 37 toys. How many toys did Bill and Lisa collect in all?

______ toys

Write Math Explain how you would find 65 + 18 by taking apart an addend.

____

____

____

Name

My Homework

Lesson 1
Take Apart Tens to Add

Homework Helper

Need help? connectED.mcgraw-hill.com

You can take apart numbers to add.

29 + 52

29 + 1 + 51

30 + 51 = 81

29 + 52

21 + 8 + 52

21 + 60 = 81

Helpful Hint
Numbers ending in zero are easier to add. Try to take apart an addend to make tens. Then add.

So, 29 + 52 = 81.

Practice

Take apart an addend to solve.

1. 35 + 27

35 + ____ + ____

____ + ____ = ____

So, 35 + 27 = ____.

2. 48 + 23

48 + ____ + ____

____ + ____ = ____

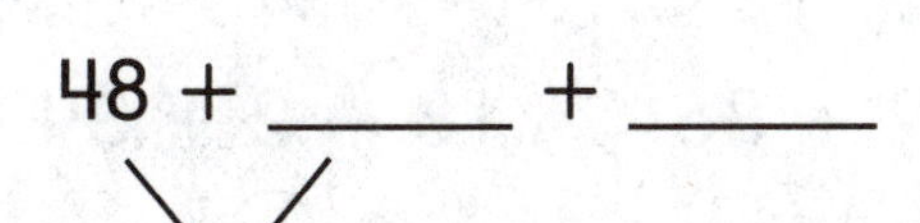

So, 48 + 23 = ____.

Take apart an addend to solve.

3. 36 + 55

36 + ______ + ______

______ + ______ = ______

So, 36 + 55 = ______.

4. 17 + 68

17 + ______ + ______

______ + ______ = ______

So, 17 + 68 = ______.

5. There were 38 winter hats and 36 scarves donated to a clothing drive. How many hats and scarves were donated in all?

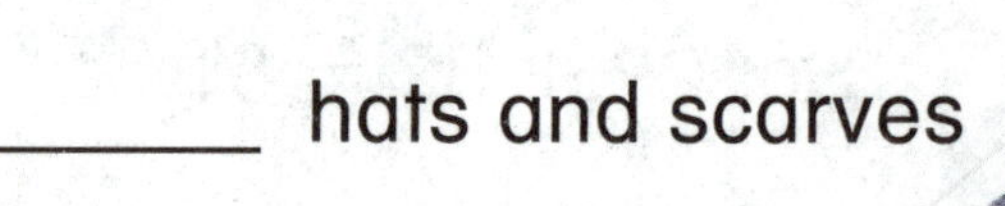

______ hats and scarves

Test Practice

6. How could you take apart 18 + 36 to find the sum?

18 + 36 = ______

9 + 9 + 36	18 + 2 + 34	1 + 8 + 36	18 + 3 + 6
○	○	○	○

Math at Home Ask your child to solve 49 + 13 by taking apart an addend.

Name ....

Regroup Ones as Tens

Lesson 2

ESSENTIAL QUESTION
How can I add two-digit numbers?

Explore and Explain

tens	ones

19 + 3 = ______

Also, ______ tens

and ______ ones.

Teacher Directions: Show the number 19 in the chart using base-ten blocks. In the ones column, below the first number, show 3 more ones cubes. Count all of the ones. Make a group of 10. Trade it for a tens rod and place the rod in the tens column. Write the sum. Write how many tens and ones.

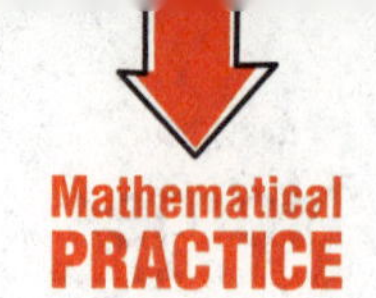

See and Show

Find 27 + 5.

Step 1
Use [ten rod] and [unit cube] to show 27 and 5.

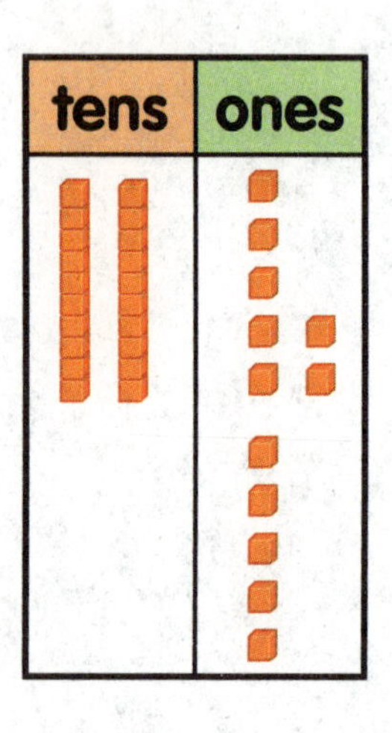

Step 2
If there are ten or more ones, **regroup** 10 ones as 1 ten.

tens	ones

Step 3
Write the number of tens and ones.

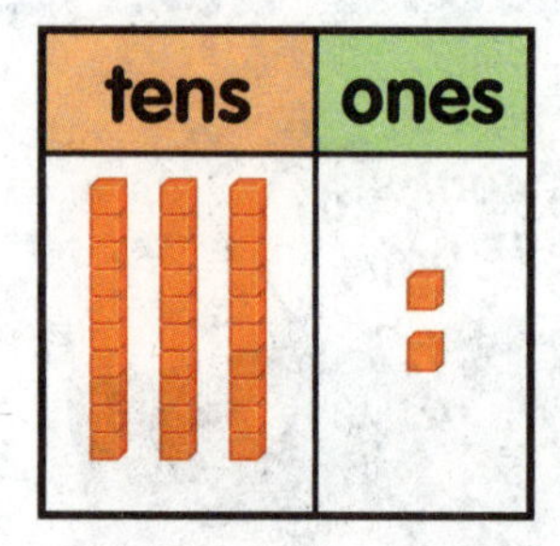

3 tens 2 ones

27 + 5 = 32

Use Work Mat 6 and base-ten blocks to add.

	Add the ones. Add the tens.	Do you regroup?	Write the sum.
1. 15 + 8	____ ten ____ ones	yes no	
2. 23 + 6	____ tens ____ ones	yes no	

How do you know if you need to regroup?

Name ....

On My Own

Use Work Mat 6 and base-ten blocks to add.

	Add the ones. Add the tens.	Do you regroup?	Write the sum.
3. 76 + 4	____ tens ____ ones	yes no	
4. 17 + 7	____ ten ____ ones	yes no	
5. 32 + 6	____ tens ____ ones	yes no	
6. 59 + 5	____ tens ____ ones	yes no	
7. 13 + 9	____ ten ____ ones	yes no	
8. 31 + 8	____ tens ____ ones	yes no	
9. 25 + 6	____ tens ____ ones	yes no	
10. 62 + 7	____ tens ____ ones	yes no	

Problem Solving

Use Work Mat 6 and base-ten blocks to solve.

11. In April, it rained 18 days. It rained 6 days in May. How many days did it rain in April and May?

______ days

12. John sold 17 bottles of water. Later, he sold 8 bottles of water. How many bottles of water did John sell in all?

______ bottles of water

13. The runners ate 6 bananas before the race. They ate 14 bananas after the race. How many bananas did the runners eat in all?

______ bananas

Explain how you can regroup 19 + 7 to find the sum.

____

____

____

Name ....

My Homework

Lesson 2
Regroup Ones as Tens

Homework Helper

Need help? connectED.mcgraw-hill.com

Find 13 + 8.

Step 1 Show each number.

Step 2 If there are 10 or more ones, regroup.

Step 3 Write the number of tens and ones.

tens	ones

13 + 8 = 21

Practice

Helpful Hint
10 ones equals 1 ten.

Circle the ones to show regrouping ten ones as 1 ten. Draw your answer.

1. 16 + 5 = _____

tens	ones

2. 18 + 4 = _____

tens	ones

Circle the ones to show regrouping ten ones as 1 ten. Draw your answer.

3. $14 + 8 =$ _____

tens	ones

4. $15 + 5 =$ _____

tens	ones

5. Max donated 16 dog collars to the dog shelter. Mia donated 6 leashes. How many items were donated?

_____ items

Vocabulary Check

Circle the picture that matches the word.

6. **regroup**

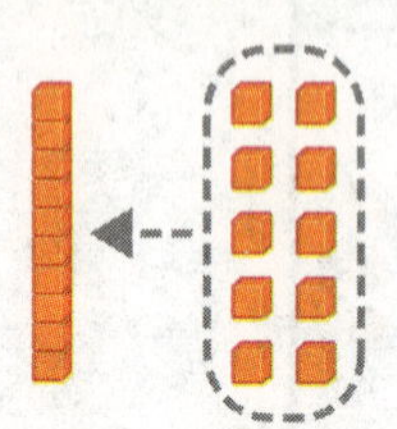

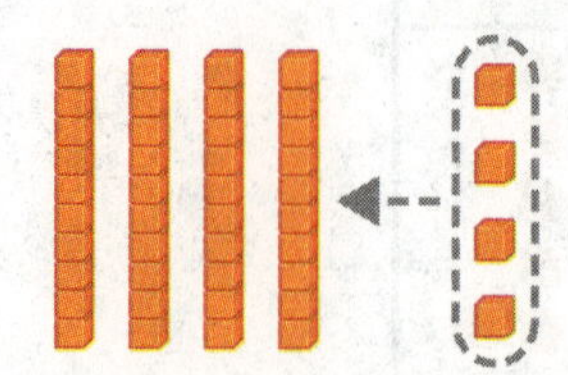

Math at Home Using toothpicks have your child show you 25 + 5. Ask how many tens there are in the answer.

Name ....

Add to a Two-Digit Number

Lesson 3

ESSENTIAL QUESTION
How can I add two-digit numbers?

Explore and Explain

tens	ones

_____ newspapers

Teacher Directions: Use base-ten blocks to show this addition story. Morgan and her friends are collecting newspapers to recycle. One group collects 7 newspapers. The other group collects 15 newspapers. How many newspapers have they collected in all? Write the number.

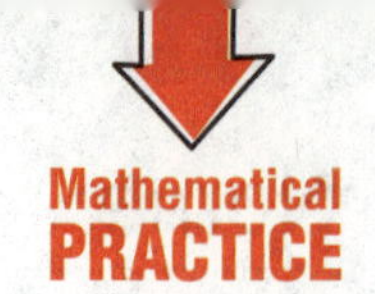

See and Show

Find 17 + 5.

Step 1
Add the ones.
7 + 5 = 12 ones

	tens	ones
	☐	
	1	7
+		5

Step 2
Regroup if needed.
Write how many.

	tens	ones
	1	
	1	7
+		5
		2

Step 3
Add the tens.
1 ten + 1 ten
= 2 tens

	tens	ones
	1	
	1	7
+		5
	2	2

Use Work Mat 6 and base-ten blocks to add.

1.

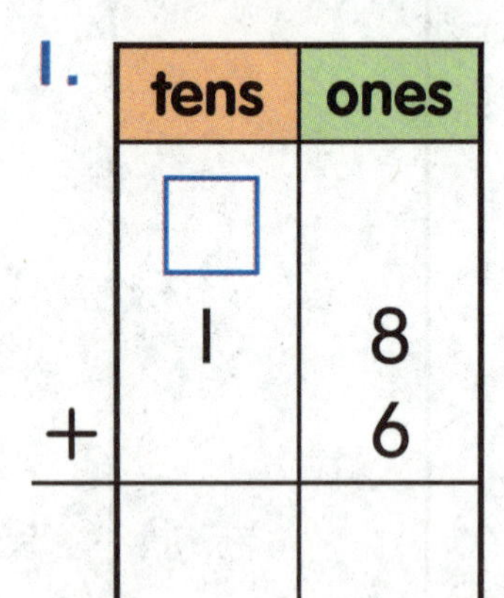

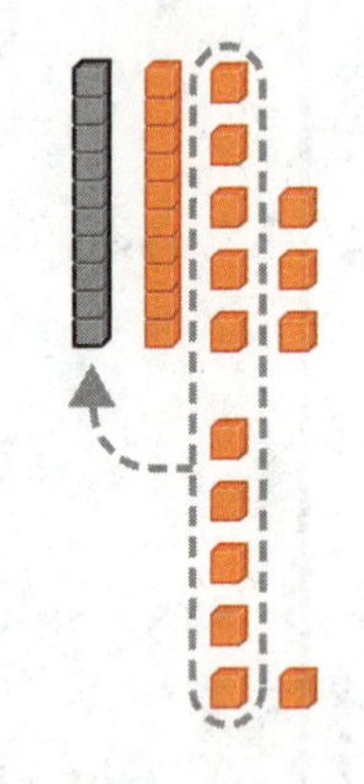

2.

	tens	ones
	☐	
	4	3
+		3

3.

	☐	
	1	9
+		6

4.

	☐	
	3	6
+		5

5.

	☐	
	1	4
+		9

Talk Math How did you show that you regrouped?

Name ....

On My Own

Use Work Mat 6 and base-ten blocks to add.

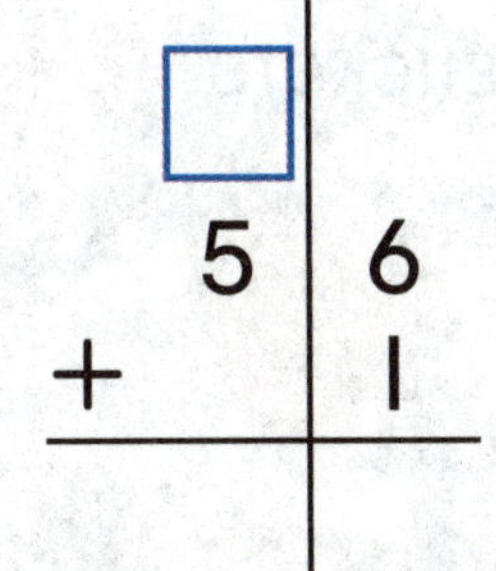

6. $\begin{array}{r} 56 \\ +\ 1 \\ \hline \end{array}$

7. $\begin{array}{r} 72 \\ +\ 8 \\ \hline \end{array}$

8. $\begin{array}{r} 38 \\ +\ 8 \\ \hline \end{array}$

9. $\begin{array}{r} 24 \\ +\ 4 \\ \hline \end{array}$

10. $\begin{array}{r} 43 \\ +\ 9 \\ \hline \end{array}$

11. $\begin{array}{r} 13 \\ +\ 7 \\ \hline \end{array}$

12. $\begin{array}{r} 51 \\ +\ 9 \\ \hline \end{array}$

13. $\begin{array}{r} 17 \\ +\ 6 \\ \hline \end{array}$

14. $\begin{array}{r} 38 \\ +\ 4 \\ \hline \end{array}$

15. $\begin{array}{r} 33 \\ +\ 7 \\ \hline \end{array}$

16. $\begin{array}{r} 22 \\ +\ 9 \\ \hline \end{array}$

17. $\begin{array}{r} 68 \\ +\ 3 \\ \hline \end{array}$

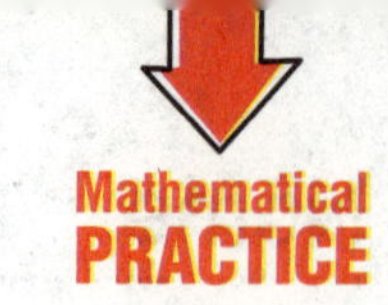

Real World Problem Solving

18. Jaya has 25 stickers on her paper. She adds 4 more stickers. How many stickers does Jaya have now?

_____ stickers

19. Micah does 19 cartwheels. Then he does 8 more. How many cartwheels does Micah do altogether?

_____ cartwheels

20. 13 children are waiting for the bus. 5 more children get in line for the bus. How many children are waiting for the bus now?

_____ children

Write Math Explain how you can solve $14 + 7$.

____

____

____

Name

My Homework

Lesson 3
Add to a Two-Digit Number

Homework Helper

Need help? connectED.mcgraw-hill.com

Find 56 + 8.

Step 1 Add the ones.
Step 2 Regroup if needed.
Step 3 Add the tens.

	tens	ones
	1	
	5	6
+		8
	6	4

So, 56 + 8 = 64.

Practice

Add.

1.

	tens	ones
	□	
	6	6
+		6

2.

	tens	ones
	□	
	8	4
+		3

3.

	tens	ones
	□	
	2	9
+		4

4.

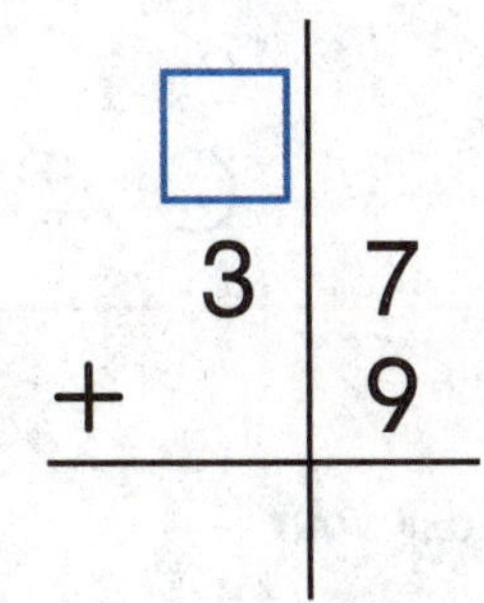

	tens	ones
	□	
	3	7
+		9

5.

	tens	ones
	□	
	2	1
+		9

6.

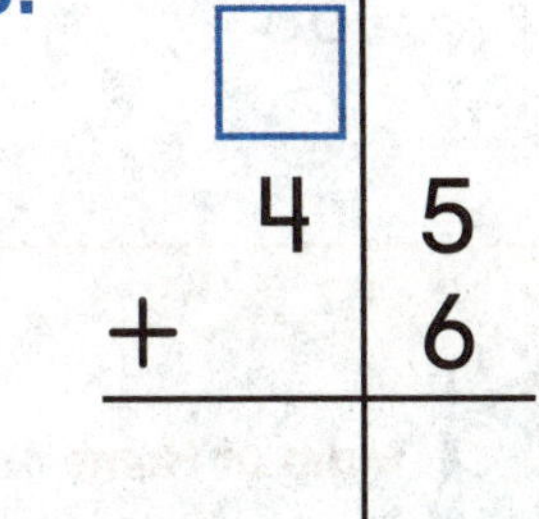

	tens	ones
	□	
	4	5
+		6

Add.

7.

	□	
	3	8
+		3

8.

	□	
	7	8
+		7

9.

	□	
	8	3
+		8

10. There are 24 students in Ms. Ito's class. 5 more students join her class. How many students are there now?

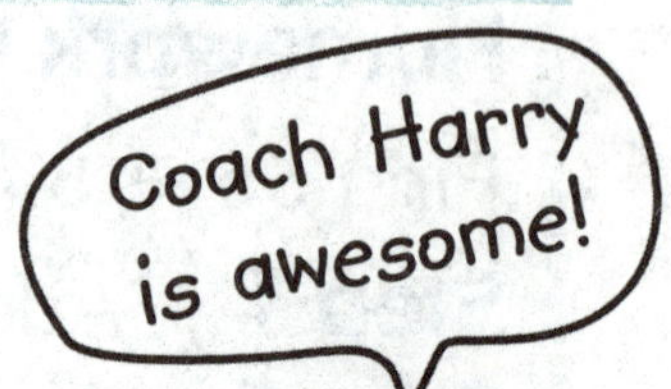

_______ students

11. Harry coaches 25 children on Wednesday. On Saturday, he coaches 9 more children. How many children does he coach in all?

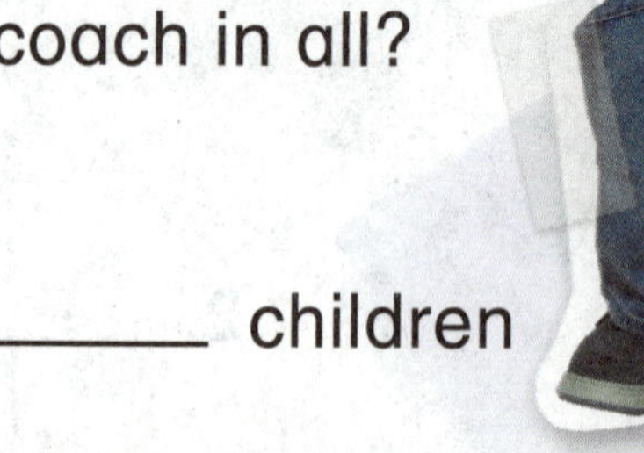

_______ children

Test Practice

12. There are 54 people that arrive early for the Tigers' game. 9 people arrive late. How many people came to the game?

63	52	44	20
○	○	○	○

Math at Home Ask your child to show you how to add 14 and 8. Then ask your child to show you how to add 27 and 2.

Name ____________________

Check My Progress

Vocabulary Check 

Complete each sentence.

sum **add** **regroup** **addends**

1. You can ________ 10 ones into 1 ten.

2. The ________ is the answer to an addition problem.

3. You ________ by joining two numbers together.

4. The numbers you join when adding are called ________.

Concept Check

Take apart an addend to solve.

5. 26 + 15

26 + ____ + ____

____ + ____ = ____

So, 26 + 15 = ____.

6. 18 + 54

18 + ____ + ____

____ + ____ = ____

So, 18 + 54 = ____.

Add.

7.

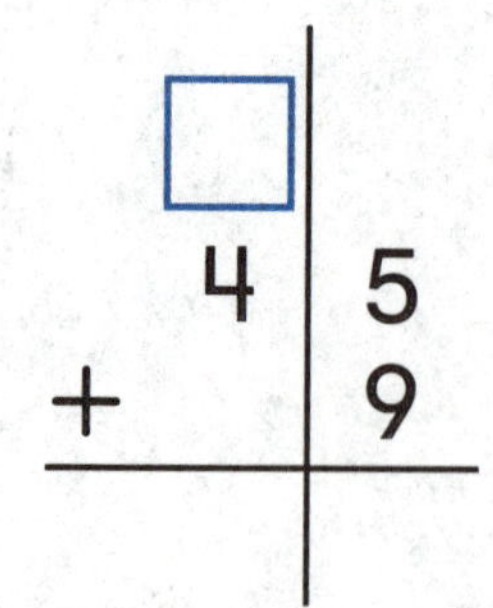

8.

	7	3
+		8

9.

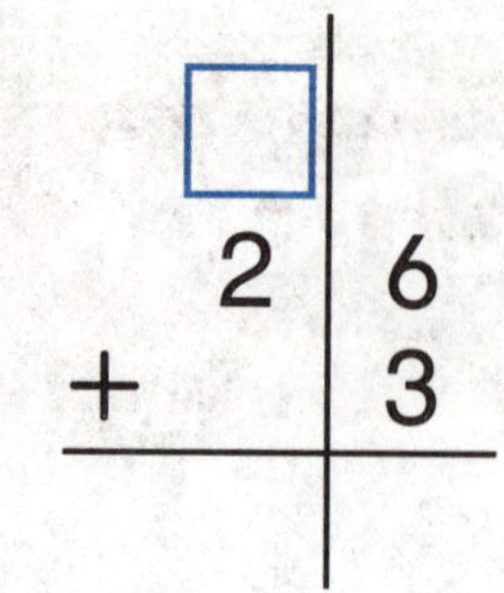

10. 24 + 9

11. 43 + 7

12. 18 + 4

13. Liam's boy scout troop collected 18 bins of plastic bottles to recycle. Landon's troop collected 9 bins. How many bins did the boy scouts collect in all?

_____ bins

Test Practice

14. 31 people are in the store. 9 more people come into the store. How many people are in the store now?

40 ○ 31 ○ 22 ○ 9 ○

Name

Add Two-Digit Numbers

Lesson 4

ESSENTIAL QUESTION
How can I add two-digit numbers?

Explore and Explain

Week 1	Week 2

_____  + _____ = _____

Teacher Directions: Model this addition story with base-ten blocks. Mr. Kay's class is selling flowers. The first week they sold 45 flowers. The second week they sold 26 flowers. How many flowers did they sell in all? Write the number sentence.

See and Show

Find 18 + 25.

Step 1
Add the ones.
Regroup if needed.

	tens	ones
	1	
	1	8
+	2	5
		3

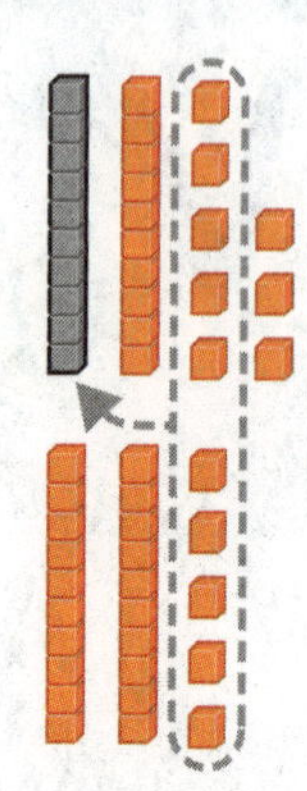

Step 2
Add the tens.

	tens	ones
	1	
	1	8
+	2	5
	4	3

So, 18 + 25 = 43

Use Work Mat 6 and base-ten blocks to add.

1.

	tens	ones
	☐	
	4	7
+	2	9

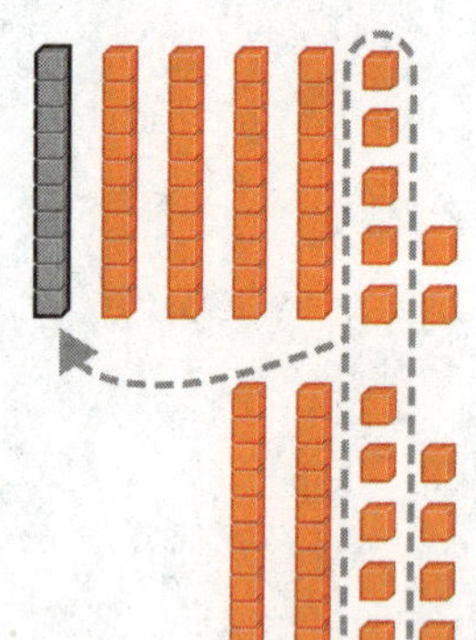

2.

	tens	ones
	☐	
	1	2
+	2	4

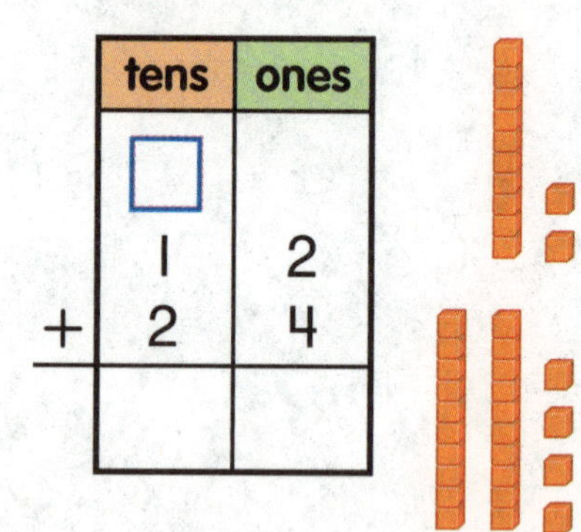

3.

	tens	ones
	☐	
	8	2
+	1	4

4.

	tens	ones
	☐	
	1	5
+	1	9

5.

	tens	ones
	☐	
	2	4
+	1	8

Talk Math What do you do first when you add two-digit numbers?

Name

On My Own

Use Work Mat 6 and base-ten blocks to add.

6.

tens	ones
□	
3	3
+ 1	8

7.

tens	ones
□	
6	4
+ 2	6

8.

tens	ones
□	
5	8
+ 1	2

9. □ 36 + 27

10. □ 22 + 10

11. □ 59 + 13

12. □ 46 + 26

13. □ 28 + 18

14. □ 9 + 32

15. 42 + 42

16. 7 + 27

17. 8 + 11

Problem Solving

18. Tia fed her turtle 30 times in June and 31 times in July. How many times did she feed her turtle?

_______ times

19. There are 25 chairs in one classroom. There are 28 chairs in another classroom. How many chairs are there in all?

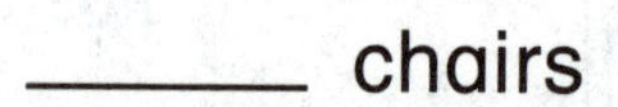

_______ chairs

20. The pet store has 14 small lizards. There are also 16 large lizards. How many lizards are at the pet store in all?

_______ lizards

How is adding 21 + 9 different than adding 21 + 19?

____

____

____

Name

My Homework

Lesson 4
Add Two-Digit Numbers

Homework Helper

Need help? connectED.mcgraw-hill.com

Find 25 + 17.

Step 1 Add the ones.
Regroup if needed.

Step 2 Add the tens.

So, 25 + 17 = 42.

	tens	ones
	1	
	2	5
+	1	7
	4	2

Practice

Add.

1.

	tens	ones
	□	
	3	5
+	2	5

2.

	tens	ones
	□	
	4	8
+	1	8

3.

	tens	ones
	□	
	1	5
+	5	9

4.

	□	
	5	6
+	3	5

5.

	□	
	2	7
+	2	8

6.

	□	
	1	4
+	3	3

Add.

7. $\begin{array}{r} 53 \\ +\ 35 \\ \hline \end{array}$

8. $\begin{array}{r} 23 \\ +\ 58 \\ \hline \end{array}$

9. $\begin{array}{r} 83 \\ +\ 13 \\ \hline \end{array}$

10. $\begin{array}{r} 62 \\ +\ 28 \\ \hline \end{array}$

11. $\begin{array}{r} 53 \\ +\ 39 \\ \hline \end{array}$

12. $\begin{array}{r} 49 \\ +\ 24 \\ \hline \end{array}$

13. Brad's family picks 22 pounds of cherries. Seth's family picks 43 pounds. How many pounds of cherries did they pick in all?

_______ pounds

Test Practice

14. 46 people came to the museum early. 39 more people came late. How many people came to the museum in all?

86	85	39	7
○	○	○	○

Math at Home Take two 2-digit numbers from your phone number and have your child add them. Example: 555-1234; 12 + 34 = 46

Name ______________________________

Rewrite Two-Digit Addition

Lesson 5

ESSENTIAL QUESTION
How can I add two-digit numbers?

Explore and Explain 

11 + 8

	tens	ones
+		

______ runs

That's Emily.

Teacher Directions: Emily's baseball team is playing a baseball game to raise money for a children's hospital. Emily's team scored 11 runs in the game. The other team scored 8 runs. How many runs were scored in all? Write the numbers in the place-value chart and add.

See and Show

You can rewrite a problem to add.

Find 35 + 26.

Step 1 Write one addend below the other addend. Line up the tens and the ones.

Step 2 Add. Regroup if necessary.

	1	
	3	5
+	2	6
	6	1

Rewrite the problem. Add.

1. 64 + 22

2. 26 + 65

3. 36 + 36

4. 73 + 19

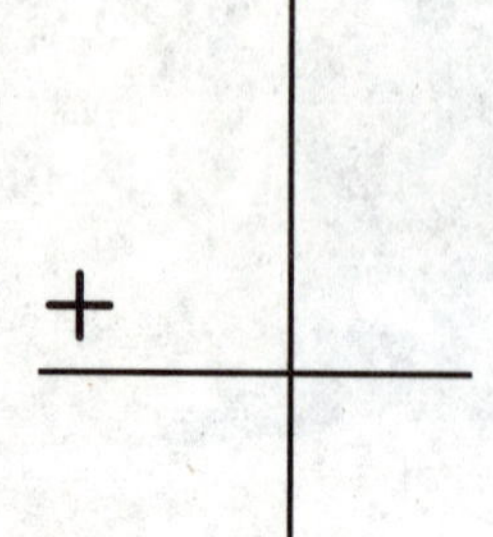

5. 47 + 18

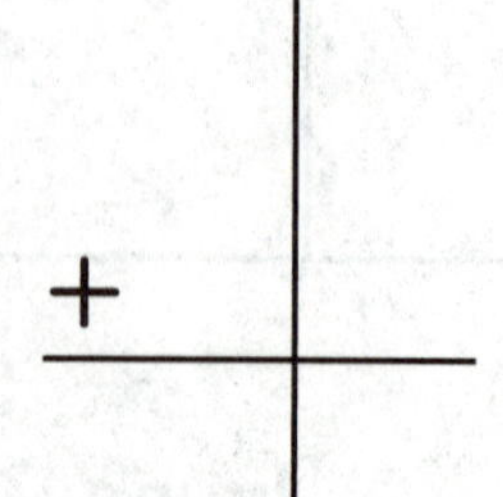

6. 56 + 37

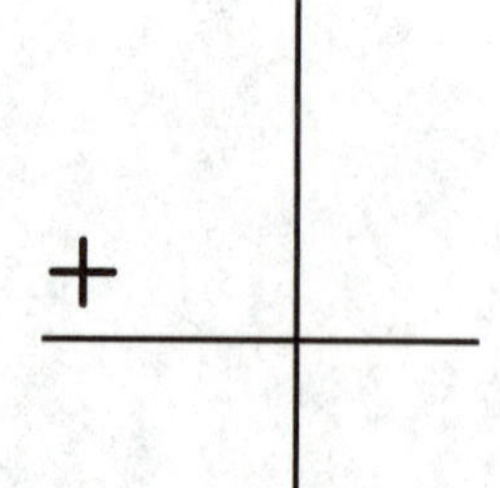

Talk Math Why is it helpful to rewrite addition?

Name ....

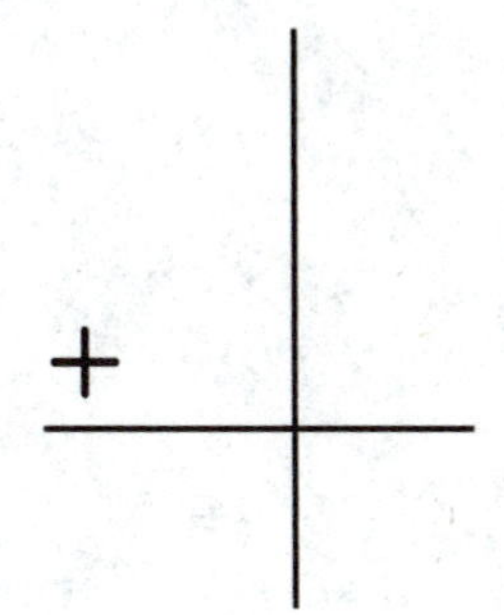

On My Own

Rewrite the problem. Add.

7. 26 + 36

+

8. 17 + 19

+

9. 74 + 16

+

10. 18 + 63

+

11. 64 + 27

+

12. 73 + 18

+

13. 56 + 23

+

14. 37 + 39

+

15. 28 + 33

+

Problem Solving

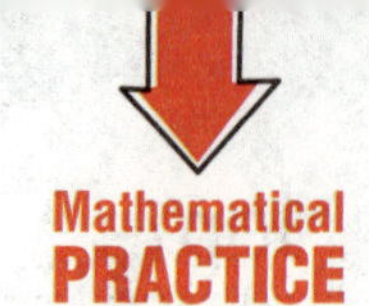

16. 38 hot dogs were sold during the baseball game. 49 hamburgers were sold. How many hot dogs and hamburgers were sold in all?

_______ hot dogs and hamburgers

17. 65 boys and 29 girls came to watch the baseball game. How many boys and girls came to watch the game?

_______ boys and girls

HOT Problem There are 13 players on the Panthers baseball team, 15 players on the Eagles team, and 14 players on the Lions team. How many players are on all three teams? Explain.

____

____

____

Name ....

My Homework

Lesson 5
Rewrite Two-Digit Addition

Homework Helper

Need help? connectED.mcgraw-hill.com

Rewrite 36 + 49 to add.

Step 1 Write one addend below the other addend.

Step 2 Add. Regroup if necessary.

	1	
	3	6
+	4	9
	8	5

Helpful Hint
Line up the ones and the tens.

Practice

Rewrite the problem. Add.

1. 64 + 15

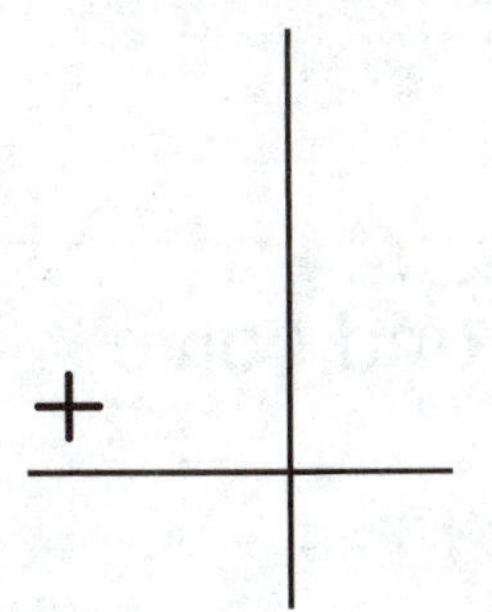

2. 26 + 57

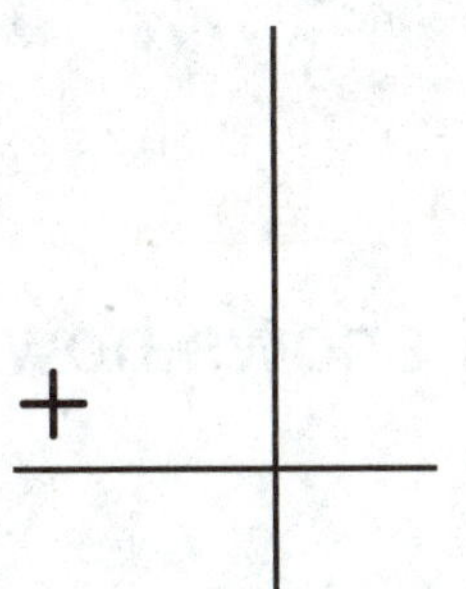

3. 61 + 28

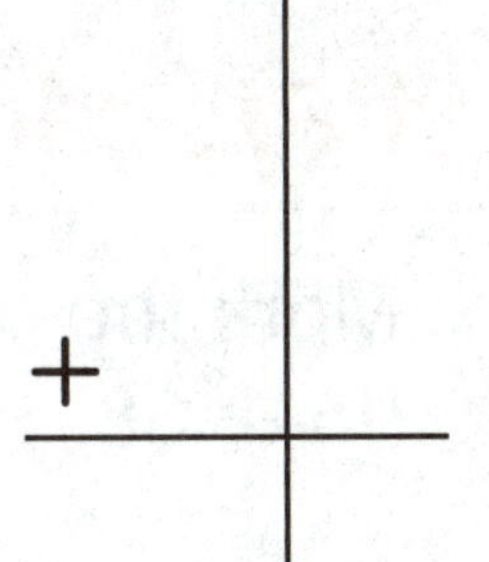

4. 37 + 47

+

5. 49 + 38

+

6. 34 + 18

+

Rewrite the problem. Add.

7. 72 + 17

+
______

8. 19 + 18

+
______

9. 29 + 35

+
______

10. 43 girls ran the race. 38 boys ran the race. How many people ran the race in all?

______ people

Test Practice

11. Mark the answer that shows how to rewrite and solve 46 + 38.

46 +38 74	316 ~~46~~ −38 8	1 46 +38 84	1 46 +38 85
○	○	○	○

Math at Home Write 35 + 38 on a piece of paper. Have your child rewrite the problem and solve it.

Name

Add Three and Four Two-Digit Numbers

Lesson 6

ESSENTIAL QUESTION
How can I add two-digit numbers?

Explore and Explain

CANNONBALL!

tens	ones

_____ cars

Teacher Directions: Miley's class is having a car wash. They washed 13 blue cars, 22 red cars, and 14 white cars. How many cars did they wash in all? Use base-ten blocks to show each number. Trace the blocks. Write the sum.

See and Show

You can add three and four two-digit numbers. Line up the ones and the tens. Add the ones first, then add the tens.

Helpful Hint
Remember, when you add the ones, look for a fact you know.

Add three numbers.

$$\begin{array}{r} 36 \\ 14 \\ +\ 24 \\ \hline 74 \end{array}$$

$6 + 4 = 10$

$10 + 4 = 14$

Add four numbers.

$$\begin{array}{r} 12 \\ 35 \\ 14 \\ +\ 24 \\ \hline 85 \end{array}$$

$4 + 4 = 8$

$8 + 5 = 13$

$13 + 2 = 15$

Add.

1. $\begin{array}{r} 17 \\ 23 \\ +\ 14 \\ \hline \end{array}$

2. $\begin{array}{r} 22 \\ 12 \\ +\ 15 \\ \hline \end{array}$

3. $\begin{array}{r} 31 \\ 19 \\ +\ 25 \\ \hline \end{array}$

4. $\begin{array}{r} 18 \\ 10 \\ +\ 32 \\ \hline \end{array}$

5. $\begin{array}{r} 12 \\ 19 \\ 21 \\ +\ 17 \\ \hline \end{array}$

6. $\begin{array}{r} 13 \\ 22 \\ 42 \\ +\ 16 \\ \hline \end{array}$

7. $\begin{array}{r} 11 \\ 25 \\ 33 \\ +\ 13 \\ \hline \end{array}$

8. $\begin{array}{r} 13 \\ 42 \\ 8 \\ +\ 36 \\ \hline \end{array}$

Talk Math How is adding three two-digit numbers like adding two two-digit numbers?

Name

Everyone stick together now!

CCSS

On My Own

Add.

9. 25 11 + 15	**10.** 51 12 + 32	**11.** 2 25 + 42	**12.** 13 33 + 45
13. 22 18 32 + 13	**14.** 15 31 19 + 20	**15.** 31 34 14 + 17	**16.** 11 15 46 + 15
17. 34 13 + 13	**18.** 16 27 + 36	**19.** 61 10 + 19	**20.** 32 15 + 38
21. 12 34 14 + 26	**22.** 10 43 17 + 20	**23.** 11 28 24 + 36	**24.** 22 36 14 + 23

Problem Solving

25. Logan has 11 paper clips. Amanda has 31 paper clips. Nate has 26 paper clips. How many paper clips do they have altogether?

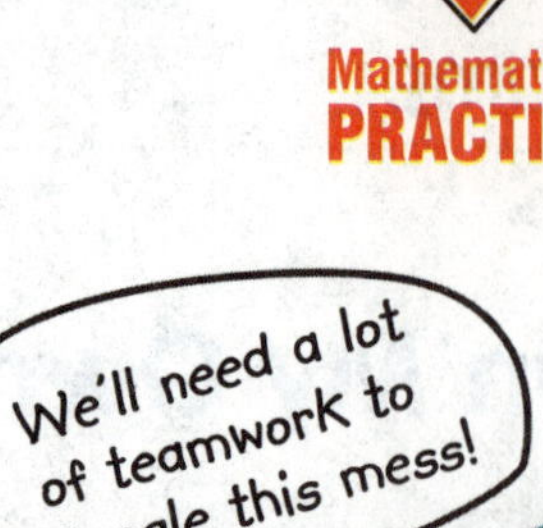

______ paper clips

26. Ben sold 28 tickets to the harvest party. Sunny sold 12 tickets. Jill sold 18 tickets. Tom sold 32 tickets. How many tickets did they sell in all?

______ tickets

HOT Problem 14 boys play basketball. 12 boys play soccer. 16 boys play baseball. How many boys play basketball, soccer, or baseball?

Jayden solved the problem like this:
$14 + 12 + 16 = 312$ boys

Tell why he is wrong. Make it right. Explain.

____

____

____

Name

My Homework

Lesson 6
Add Three and Four Two-Digit Numbers

Homework Helper

Need help? connectED.mcgraw-hill.com

To add three or four two-digit numbers, line up the ones and the tens.

Add the ones, then add the tens.

$$\begin{array}{r} 15 \\ 16 \\ +\ 46 \\ \hline 77 \end{array} \qquad \begin{array}{r} 18 \\ 21 \\ 46 \\ +\ 12 \\ \hline 97 \end{array}$$

Helpful Hint
Look for a fact you know or doubles.

Practice

Add.

1. $\begin{array}{r} 53 \\ 27 \\ +\ 10 \\ \hline \end{array}$

2. $\begin{array}{r} 52 \\ 23 \\ +\ 18 \\ \hline \end{array}$

3. $\begin{array}{r} 11 \\ 19 \\ +\ 24 \\ \hline \end{array}$

4. $\begin{array}{r} 26 \\ 35 \\ 24 \\ +\ 11 \\ \hline \end{array}$

5. $\begin{array}{r} 23 \\ 36 \\ 16 \\ +\ 11 \\ \hline \end{array}$

6. $\begin{array}{r} 23 \\ 33 \\ 13 \\ +\ 15 \\ \hline \end{array}$

Read and solve the problem.

7. 13 students play the bells. 16 students play the drums. 24 students play the recorder. 14 students play trumpet. How many students play instruments?

_______ students

8. Mr. Tan picks 24 apples. Mrs. Tan picks 35 apples. Their son picks 26 apples. How many apples do the Tans pick in all?

_______ apples

Test Practice

9. A class collects 20 toys on Monday, 15 toys on Tuesday and 10 toys on Wednesday. How many toys did they collect on Monday and Wednesday?

10	20	30	35
○	○	○	○

Math at Home Have your child explain how to add 28 + 12 + 35.

Name

Problem Solving

STRATEGY: Make a Model

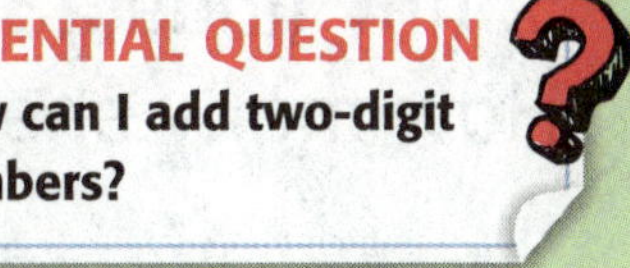

Lesson 7

ESSENTIAL QUESTION
How can I add two-digit numbers?

Tools

Mark and Dan were in a walk-a-thon to raise money for their basketball team. Mark walked 35 blocks. Dan walked 52 blocks. How many blocks did they walk in all?

1 Understand

Underline what you know.
Circle what you need to find.

2 Plan

How will I solve the problem?

3 Solve

Make a model.

87 blocks

4 Check

Is my answer reasonable? Explain.

Practice the Strategy

Sara, Mary, and Jonny are collecting shells at the beach. Sara finds 45 shells. Jonny finds 23 shells. Mary finds 15 shells. How many shells do they find?

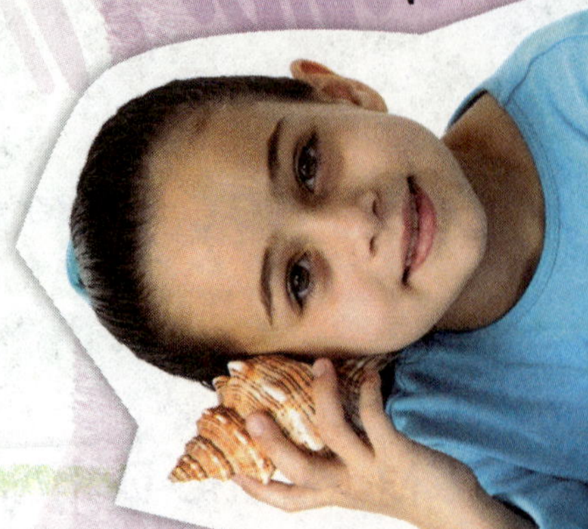

1 Understand Underline what you know. Circle what you need to find.

2 Plan How will I solve the problem?

3 Solve I will...

_______ shells

4 Check Is my answer reasonable? Explain.

CCSS

Name

Apply the Strategy

1. Karenna ran 15 blocks Tuesday.
She ran 20 blocks on Wednesday.
How many blocks did she run in all?

_____ blocks

2. Bradley got 33 action figures for his birthday.
He had 12 already. How many will he have
if he buys 15 more?

_____ action figures

3. Nora made 16 points in her basketball game.
Jenna made 24 points and Alicia made 22.
No one else scored on their team.
What was their final score?

_____ points

Review the Strategies

Choose a strategy
- Write a number sentence.
- Find a pattern.
- Make a model.

4. Melissa has 23 pairs of earrings. She buys 2 more pairs. Then she is given 6 more pairs for her birthday. How many pairs of earrings does she have in all?

_______ pairs

5. Jessica made muffins for the bake sale. She made 24 blueberry muffins and 10 strawberry muffins. She also made 12 banana muffins. How many muffins did she make in all?

_______ muffins

6. Jacob and Luis collected glass bottles to recycle. The first day they collected 12 bottles. The second day they collected 15. The third day they collected 18. If this pattern continues, how many will they collect on the fourth day?

_______ bottles

Name ....

My Homework

Lesson 7

Problem Solving: Make a Model

eHelp

Karen and Jake are counting ribbons. Jake counts 20 blue ribbons. Karen counts 16 red ribbons. How many ribbons did they count in all?

First prize in ribbon counting goes to...

1 Understand

Underline what you know.
Circle what you need to find.

2 Plan

How will I solve the problem?

3 Solve

Make a model.

$20 + 16 = 36$ ribbons

4 Check

Is my answer reasonable?

Underline what you know. Circle what you need to find. Make a model to solve.

1. Angela and Brittany count 8 windows in the first building, 5 in the next building, and 13 in the last building. How many windows did they count in all?

_______ windows

2. Eric found 14 pennies in the couch, 6 pennies in his mom's car, and 2 pennies under his bed. How many pennies did he find in all?

_______ pennies

3. The girls collected 57 bottle caps. The boys collected 42 caps. How many bottle caps did they collect in all?

_______ caps

Math at Home Ask your child to add 13 + 21 + 37. Have your child explain how he or she added the numbers.

Name

Fluency Practice

Add

1. $\begin{array}{r} 14 \\ +\ 3 \\ \hline \end{array}$	**2.** $\begin{array}{r} 16 \\ +\ 6 \\ \hline \end{array}$	**3.** $\begin{array}{r} 13 \\ +\ 8 \\ \hline \end{array}$
4. $\begin{array}{r} 14 \\ +\ 9 \\ \hline \end{array}$	**5.** $\begin{array}{r} 15 \\ +\ 6 \\ \hline \end{array}$	**6.** $\begin{array}{r} 16 \\ +\ 3 \\ \hline \end{array}$
7. $\begin{array}{r} 11 \\ +\ 9 \\ \hline \end{array}$	**8.** $\begin{array}{r} 12 \\ +\ 2 \\ \hline \end{array}$	**9.** $\begin{array}{r} 14 \\ +\ 0 \\ \hline \end{array}$
10. $\begin{array}{r} 18 \\ +\ 5 \\ \hline \end{array}$	**11.** $\begin{array}{r} 19 \\ +\ 1 \\ \hline \end{array}$	**12.** $\begin{array}{r} 17 \\ +\ 2 \\ \hline \end{array}$

Fluency Practice

Add

1. 13 + 6

2. 16 + 2

3. 13 + 8

4. 26 + 5

5. 13 + 1

6. 13 + 3

7. 21 + 9

8. 16 + 2

9. 11 + 0

10. 13 + 2

11. 13 + 4

12. 12 + 7

Together, we make beautiful music!

Name

My Review

Chapter 3

Add Two-Digit Numbers

Vocabulary Check

Draw lines to match.

1. regroup — $3 + 4 = 7$

2. sum —

$$\begin{array}{r} \boxed{1} \\ 34 \\ +\ 17 \\ \hline 51 \end{array}$$

3. add — $8 + 1 = 9$ ↑

Concept Check

Take apart an addend to solve.

4. 33 + 58

_____ + _____ + 58

_____ + _____ = _____

So, 33 + 58 = _____.

5. 17 + 46

17 + _____ + _____

_____ + _____ = _____

So, 17 + 46 = _____.

Add.

6.
```
   3 | 4
+  2 | 7
```

7.
```
   4 | 3
+  2 | 8
```

8.
```
   1 | 5
+  3 | 7
```

9.
```
   1 | 1
+  1 | 8
```

10.
```
   1 | 5
+  7 | 3
```

11.
```
   2 | 8
+  6 | 3
```

12.
```
   36
+  36
```

13.
```
   26
+  45
```

14.
```
   32
+  16
```

15.
```
   23
   17
+  30
```

16.
```
   52
   22
+  20
```

17.
```
   16
   12
+  24
```

18.
```
   16
   25
   44
+  11
```

19.
```
   57
   17
   13
+  10
```

20.
```
   21
   13
   13
+  16
```

CCSS

Name ____________________

Problem Solving

21. Kaylee sold 35 raffle tickets. Chloe sold 27 and Noah sold 33. How many raffle tickets did they sell in all?

_______ tickets

22. Peyton has 13 marbles. She finds 11 more marbles. How many marbles does Peyton have in all?

We're on a roll!

_______ marbles

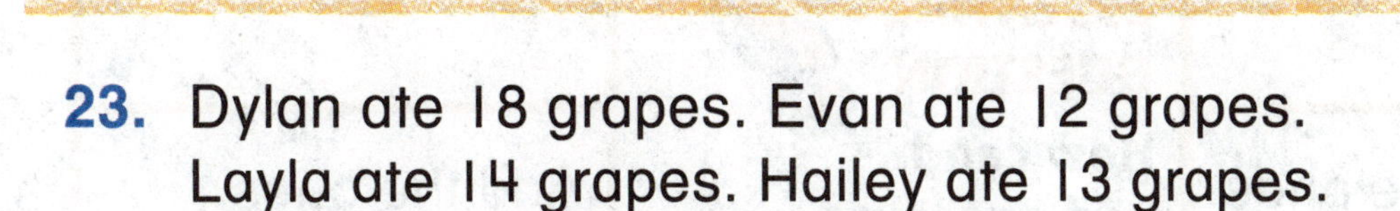

23. Dylan ate 18 grapes. Evan ate 12 grapes. Layla ate 14 grapes. Hailey ate 13 grapes. How many grapes were eaten?

_______ grapes

Test Practice

24. The red team has 20 boys, the blue team has 19 boys, the green team has 19 boys, and the yellow team has 20 boys. How many boys are there in all?

80	78	59	58
○	○	○	○

Reflect

Chapter 3

Answering the Essential Question

Show the ways you can add to solve problems.

ESSENTIAL QUESTION

How can I add two-digit numbers?

Add the ones. Regroup if needed.

$$\begin{array}{r} 17 \\ +\ 8 \\ \hline \end{array}$$

Add the ones. Regroup if needed. Add the tens.

$$\begin{array}{r} 24 \\ +\ 17 \\ \hline \end{array}$$

Add the ones. Regroup if needed. Add the tens.

$$\begin{array}{r} 18 \\ 12 \\ +\ 14 \\ \hline \end{array}$$

Add the ones. Regroup if needed. Add the tens.

$$\begin{array}{r} 24 \\ 35 \\ 10 \\ +\ 15 \\ \hline \end{array}$$

Don't stop until you reach the top!

Go team! Go!

Chapter 4

Subtract Two-Digit Numbers

ESSENTIAL QUESTION

How can I subtract two-digit numbers?

Watch a video!

My Common Core State Standards

Operations and Algebraic Thinking

2.OA.1 Use addition and subtraction within 100 to solve one-and two-step word problems involving situations of adding to, taking from, putting together, taking apart, and comparing, with unknowns in all positions.

2.NBT.5 Fluently add and subtract within 100 using strategies based on place value, properties of operations, and/or the relationship between addition and subtraction.

2.NBT.9 Explain why addition and subtraction strategies work, using place value and the properties of operations.

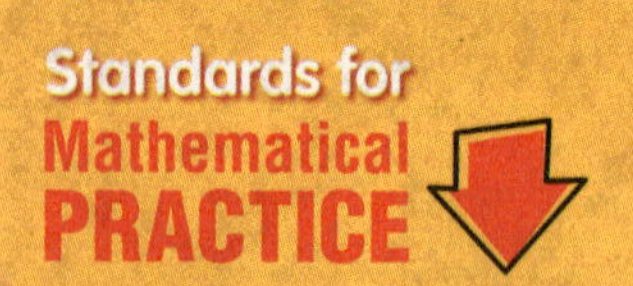

1. Make sense of problems and persevere in solving them.
2. Reason abstractly and quantitatively.
3. Construct viable arguments and critique the reasoning of others.
4. Model with mathematics.
5. Use appropriate tools strategically.
6. Attend to precision.
7. Look for and make use of structure.
8. Look for and express regularity in repeated reasoning.

= focused on in this chapter

Name ________________

Am I Ready?

Check ← Go online to take the Readiness Quiz

Write how many.

1. How many tens are in 30? ______
2. How many tens are in 70? ______
3. How many tens are in 90? ______

Subtract.

4. $9 - 6 =$ ______
5. $8 - 2 =$ ______
6. $7 - 2 =$ ______
7. $6 - 5 =$ ______
8. $4 - 3 =$ ______
9. $5 - 3 =$ ______

Write a number sentence.

10. Maggie has 10 pretzels in her lunch. Lamont eats 4 of Maggie's pretzels. How many pretzels are left?

______ ◯ ______ ◯ ______ pretzels

How Did I Do? Shade the boxes to show the problems you answered correctly.

1	2	3	4	5	6	7	8	9	10

Name

My Math Words

Review Vocabulary

difference fact family sum

Complete the addition and subtraction sentences. Complete each sentence.

$18 - 6 =$ ☐

$18 - 12 =$ ☐

$12 + 6 =$ ◯

$6 + 12 =$ ◯

The boxed numbers are called the ______________.

The circled numbers are called the ______________.

The numbers 6, 12, and 18 are called a ______________.

My Vocabulary Cards

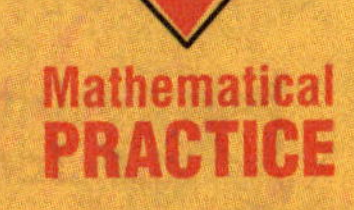

Directions:
Ideas for Use

- Have students use blank cards to write subtraction facts they know.
- Ask students to use the blank cards to write their own vocabulary cards.

Name ....

Two-Digit Fact Families

Lesson 1

ESSENTIAL QUESTION
How can I subtract two-digit numbers?

Explore and Explain

____ + ____ = ____   ____ − ____ = ____

____ + ____ = ____   ____ − ____ = ____

Teacher Directions: Use base-ten blocks to solve. Lindsay has 16 snowballs. She throws 7 of them. How many are left? Write the three numbers in the fact family. Then write the related number sentences.

See and Show

The numbers in this addition and subtraction fact family are 18, 15, and 3.

15 + 3 = 18　　18 − 15 = 3

3 + 15 = 18　　18 − 3 = 15

Complete each fact family.

1.

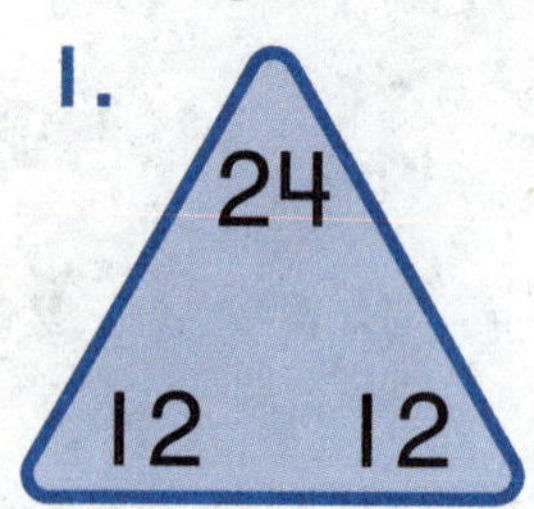

12 + 12 = ____　　24 − 12 = ____

2.

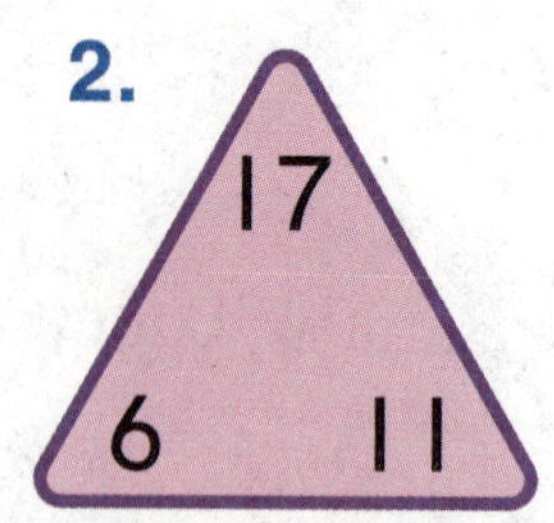

11 + 6 = ____　　17 − 11 = ____

6 + 11 = ____　　17 − 6 = ____

3.

16 + 4 = ____　　20 − 4 = ____

____ + ____ = ____　　____ − ____ = ____

Talk Math How are these exercises like fact families that you have learned earlier?

Name ______________________

CCSS

On My Own

Complete each fact family.

4.

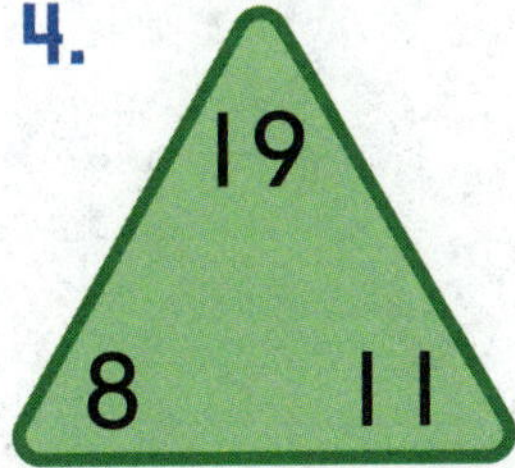

___ + ___ = ___ ___ − ___ = ___

___ + ___ = ___ ___ − ___ = ___

5.

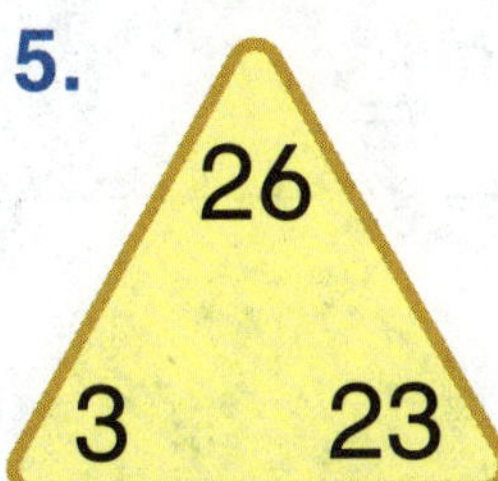

___ + ___ = ___ ___ − ___ = ___

___ + ___ = ___ ___ − ___ = ___

6.

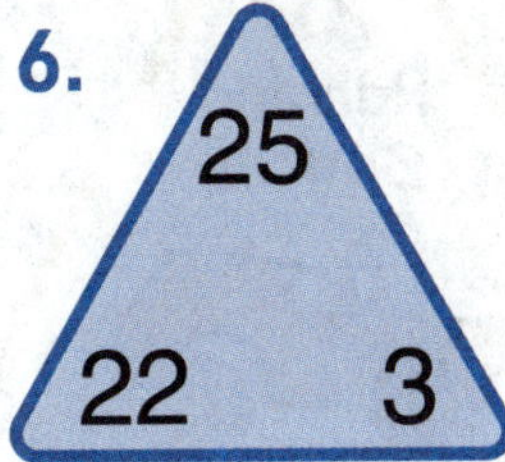

___ + ___ = ___ ___ − ___ = ___

___ + ___ = ___ ___ − ___ = ___

7.

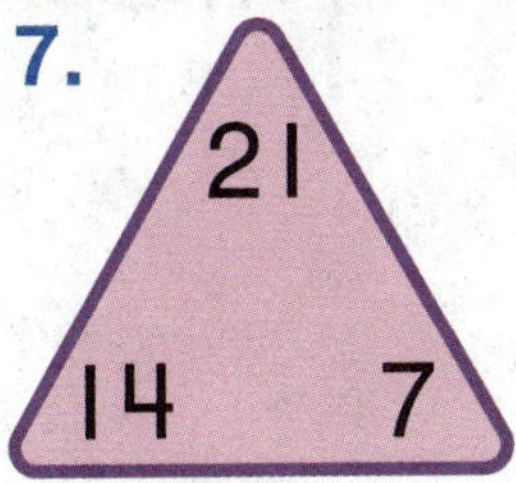

___ + ___ = ___ ___ − ___ = ___

___ + ___ = ___ ___ − ___ = ___

8.

___ + ___ = ___ ___ − ___ = ___

___ + ___ = ___ ___ − ___ = ___

Problem Solving

9. Lizzie has 12 necklaces without beads. She also has 22 necklaces that have beads on them. How many necklaces does Lizzie have? Write the fact family that can be made from these numbers.

_____ necklaces

_____ + _____ = _____ _____ − _____ = _____

_____ + _____ = _____ _____ − _____ = _____

10. How many parking spaces are open in the parking lot?

_____ parking spaces

HOT Problem Sophia wrote this fact family. Tell why Sophia is wrong. Make it right.

$76 + 12 = 88$ $76 - 12 = 88$

$12 + 76 = 88$ $12 - 76 = 88$

____

____

____

Name ....

My Homework

Lesson 1
Two-Digit Fact Families

Homework Helper

Need help? connectED.mcgraw-hill.com

The numbers 15, 11, and 4 make up the numbers in a fact family.

$11 + 4 = 15$ $15 - 11 = 4$

$4 + 11 = 15$ $15 - 4 = 11$

Practice

Complete each fact family.

1.

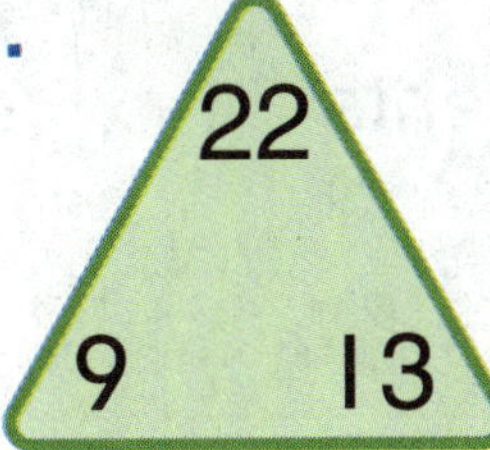

___ + ___ = ___ ___ − ___ = ___

___ + ___ = ___ ___ − ___ = ___

2.

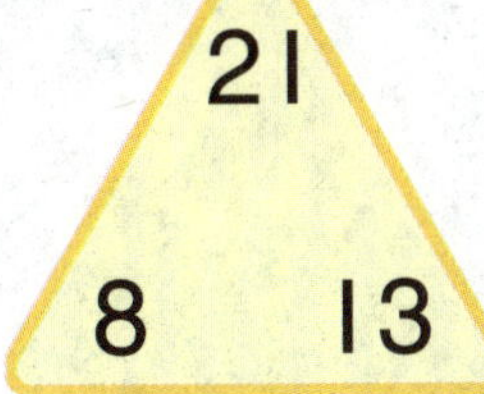

___ + ___ = ___ ___ − ___ = ___

___ + ___ = ___ ___ − ___ = ___

3.

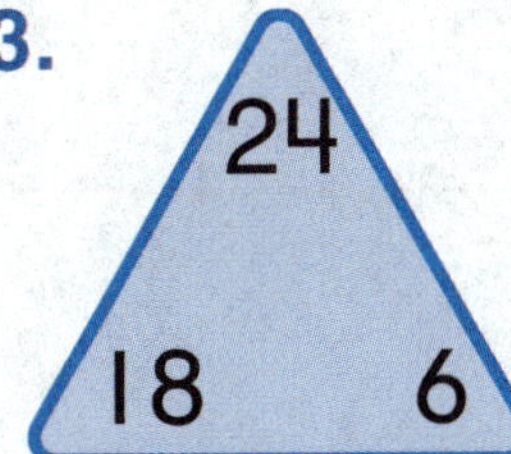

___ + ___ = ___ ___ − ___ = ___

___ + ___ = ___ ___ − ___ = ___

Complete the fact family.

4. 

____ + ____ = ____   ____ − ____ = ____

Use what you know about fact families to solve.

5. Marcus counts 24 children playing. 11 of the children are boys. How many girls are playing?

______ girls

6. There are 49 second graders. 26 of them are in Miss Johnson's class. How many students are in Mrs. Stewart's class?

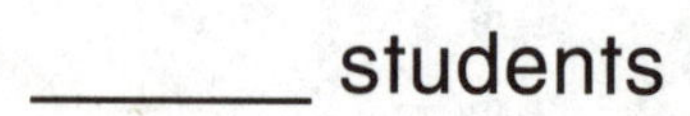

______ students

Test Practice

7. Which number sentence does not belong in the fact family for the numbers 7, 18, and 25?

18 + 7 = 25 ○	18 − 25 = 7 ○
7 + 18 = 25 	25 − 7 = 18 

Math at Home Have your child write the fact family for the numbers 22, 13, and 35.

Operations and Algebraic Thinking
2.OA.1
CCSS

Name

Take Apart Tens to Subtract

Lesson 2

ESSENTIAL QUESTION
How can I subtract two-digit numbers?

Explore and Explain

22 = _____ tens _____ ones

56 − _____ = _____

_____ − _____ = _____

_____ leaves

Teacher Directions: There are 56 leaves on a tree. 22 leaves fall off of the tree. How many leaves are left? Use base-ten blocks to show 56 leaves. Take apart 22 into tens and ones. Subtract the tens from 56. Then subtract the ones from that difference.

See and Show

You can take apart numbers to subtract.

To find 35 − 13, take apart a number to make a ten. Then subtract.

35 − 13

10 ___ 3 ___

35 − 10 = 25

25 − 3 = 22

So, 35 − 13 = 22.

Helpful Hint
Take apart 13 as 10 and 3.

Take apart a number to make a ten. Then subtract.

1. 64 − 12

_____ _____

64 − _____ = _____

_____ − _____ = _____

So, 64 − 12 = _____.

2. 45 − 24

_____ _____

45 − _____ = _____

_____ − _____ = _____

So, 45 − 24 = _____.

Talk Math Explain how you decide what to subtract when working these problems.

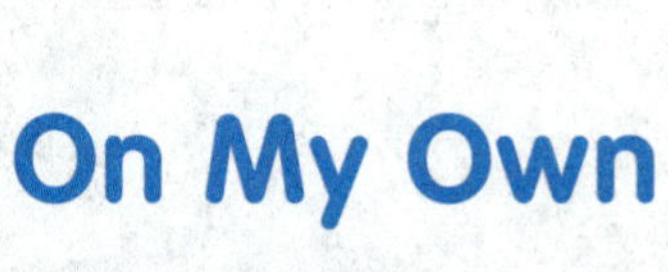

Name ______________________

On My Own

Take apart a number to make a ten. Then subtract.

3. 47 − 26

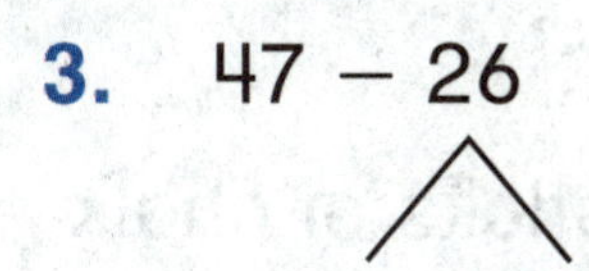

____ ____

47 − ____ = ____

____ − ____ = ____

So, 47 − 26 = ____.

4. 77 − 43

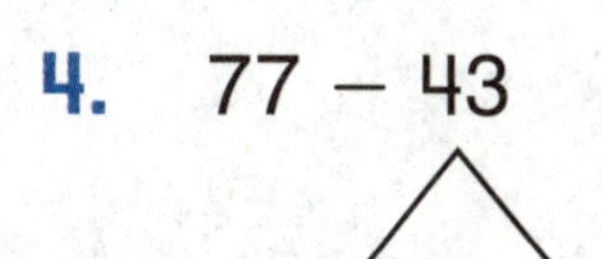

____ ____

77 − ____ = ____

____ − ____ = ____

So, 77 − 43 = ____.

5. 85 − 32

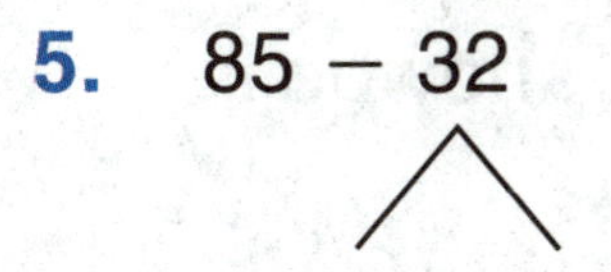

____ ____

85 − ____ = ____

____ − ____ = ____

So, 85 − 32 = ____.

6. 75 − 55

____ ____

75 − ____ = ____

____ − ____ = ____

So, 75 − 55 = ____.

Problem Solving

Mathematical PRACTICE

7. Cara has 28 sticks of sidewalk chalk. She lets her friend borrow 11 of them. How many sticks of chalk does Cara have now?

_____ sticks of chalk

8. Shawna had 37 snowballs. She threw 23 of them. How many snowballs does Shawna have left?

_____ snowballs

HOT Problem There are 22 students in Ms. Marshall's class. 8 students like science, 2 like art, and 3 like music. The rest of the students like math. How many students like math?

____

____

____

Name ______________________________

My Homework

Lesson 2
Take Apart Tens to Subtract

Homework Helper

Need help? connectED.mcgraw-hill.com

To find 48 − 15, take apart a number to make a ten.
Then subtract.

48 − 15

10   5

48 − 10 = 38
38 − 5 = 33

So, 48 − 15 = 33.

Helpful Hint
Take apart 15 as 10 and 5.

Practice

Take apart a number to make a ten. Then subtract.

1. 54 − 13

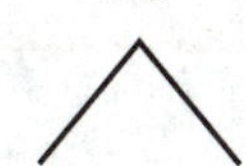

______ ______

54 − ______ = ______

______ − ______ = ______

So, 54 − 13 = ______.

2. 35 − 14

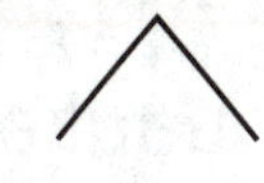

______ ______

35 − ______ = ______

______ − ______ = ______

So, 35 − 14 = ______.

3. There are 87 insects in a jar. Someone leaves the lid off and some crawl away. There are 54 insects left in the jar. How many crawled away?

_______ insects

4. Emily has 25 onions and 14 peppers that she grew in her garden. How many more onions does she have than peppers?

_______ onions

5. There were 25 days of rain in April. It rained 16 days in May. If there are 61 days total in April and May, how many days did it not rain?

_______ days

Test Practice

6. How would you take apart 14 to solve 28 − 14?

7 and 7 ○	12 and 2 ○
10 and 4 ○	20 and 8 ○

Math at Home Have your child take apart 16 to make a ten to find 87 − 16.

Name ____________________

Operations and Algebraic Thinking

2.OA.1, 2.NBT.5, 2.NBT.9

CCSS

Regroup a Ten as Ones

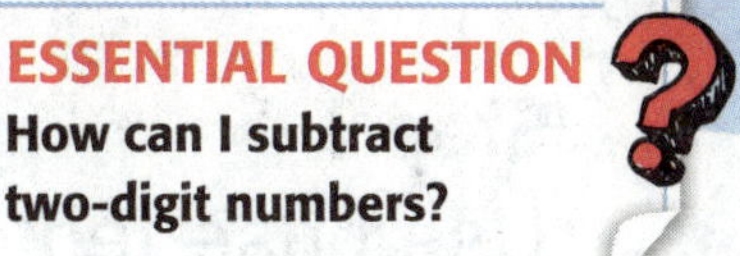

Lesson 3

ESSENTIAL QUESTION

How can I subtract two-digit numbers?

Explore and Explain

tens	ones

______ sea shells

Teacher Directions: Adam has 23 sea shells. Use base-ten blocks to show the sea shells. He gives 5 of them away. Regroup and take away that many blocks. How many sea shells are left? Draw the base-ten blocks that are left in the chart. Write how many sea shells are left.

See and Show

Find 24 – 8.

Step 1
Use base-ten blocks to show 24.

tens | ones

2 tens 4 ones

Step 2
Subtract the ones. There are not enough ones to subtract. Regroup.

tens | ones

1 ten 14 ones

Step 3
Subtract 8 ones.

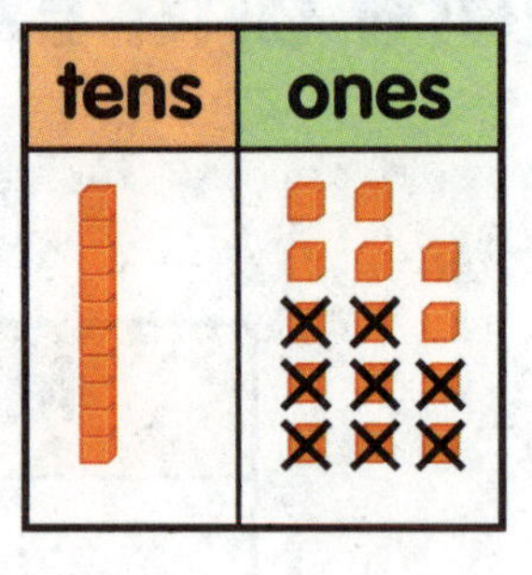

1 ten 6 ones

24 – 8 = 16

Use Work Mat 6 and base-ten blocks to subtract.

	Subtract the ones. Do you need to regroup?	Write the difference.
1. 31 – 4	no yes	31 – 4 = ______
2. 27 – 5	no yes	27 – 5 = ______

How do you know when you need to regroup? Explain.

Name ______________________________

On My Own

Helpful Hint
If there are not enough ones to subtract, regroup 1 ten as 10 ones.

Use Work Mat 6 and base-ten blocks to subtract.

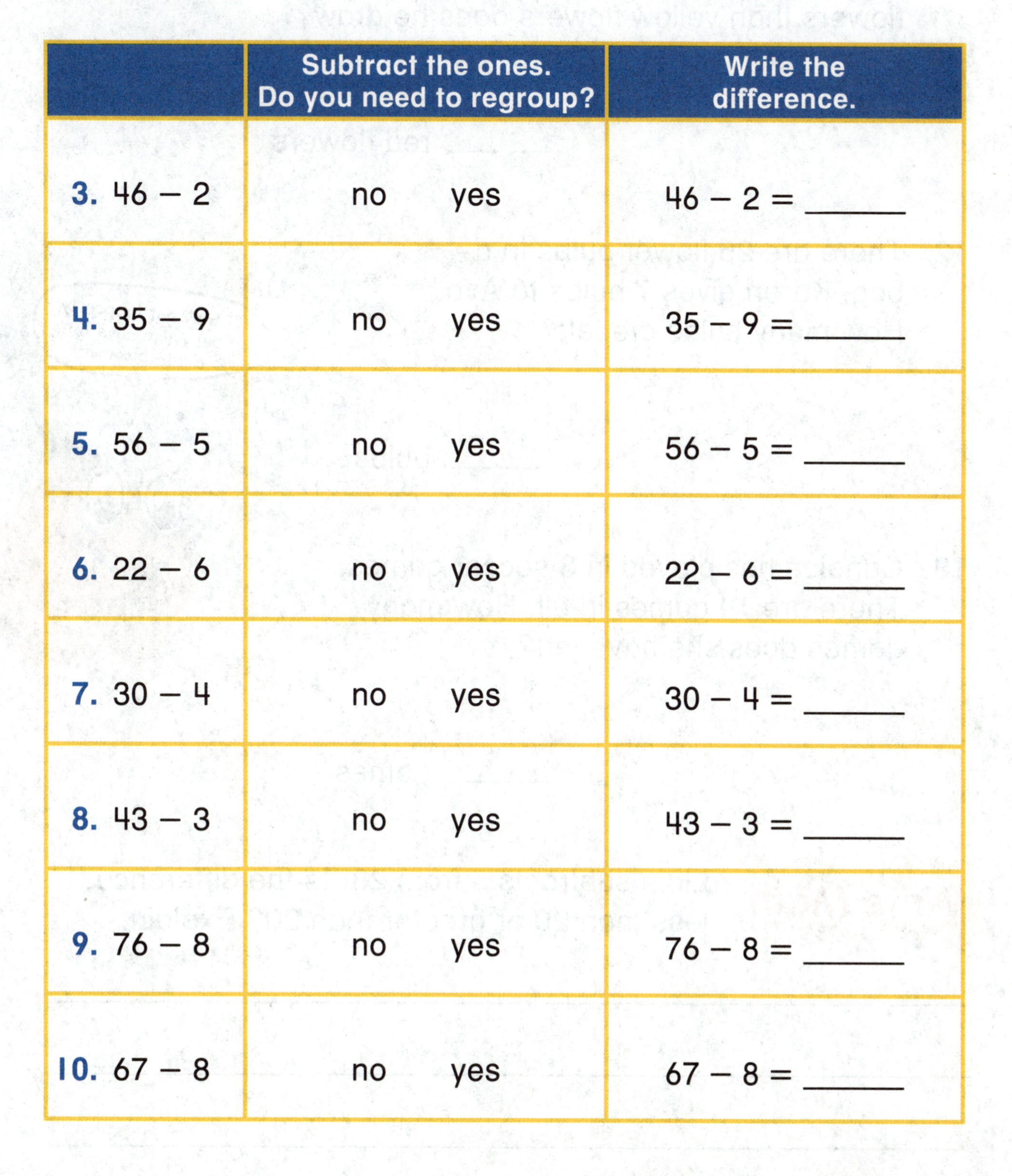

	Subtract the ones. Do you need to regroup?	Write the difference.
3. 46 − 2	no yes	46 − 2 = ______
4. 35 − 9	no yes	35 − 9 = ______
5. 56 − 5	no yes	56 − 5 = ______
6. 22 − 6	no yes	22 − 6 = ______
7. 30 − 4	no yes	30 − 4 = ______
8. 43 − 3	no yes	43 − 3 = ______
9. 76 − 8	no yes	76 − 8 = ______
10. 67 − 8	no yes	67 − 8 = ______

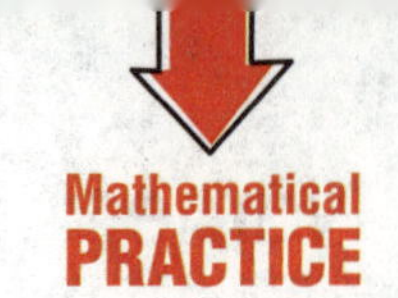

Problem Solving

Use Work Mat 6 and base-ten blocks to subtract.

11. Logan draws 23 red flowers. He draws 14 yellow flowers. How many more red flowers than yellow flowers does he draw?

_______ red flowers

12. There are 25 flower bulbs in a bag. Karen gives 7 bulbs to Ava. How many bulbs are left?

_______ bulbs

13. Candice has played in 8 soccer games. There are 14 games in all. How many games does she have left?

_______ games

Write Math Liam subtracts 5 from 23. Is the difference less than 20 or greater than 20? Explain.

____

____

____

Name ______________________

My Homework

Lesson 3
Regroup a Ten as Ones

Homework Helper

Need help? connectED.mcgraw-hill.com

Find 35 – 7.

Step 1
Show 35.

tens	ones

3 tens 5 ones

Step 2
Subtract the ones. There are not enough ones to subtract. Regroup.

tens	ones

2 tens 15 ones

Step 3
Subtract 7 ones.

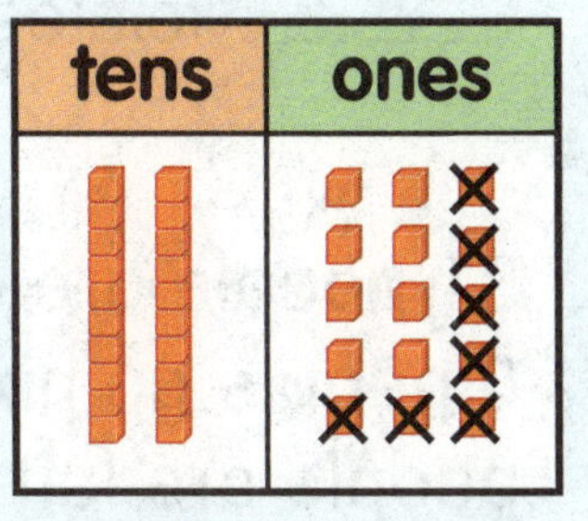

2 tens 8 ones
35 – 7 = 28

Helpful Hint
If there are not enough ones to subtract, regroup 1 ten as 10 ones.

Practice

Regroup the tens. Draw the ones. Cross out cubes to show subtraction. Write the answer.

1. 26 – 9 = ______

tens	ones

2. 23 – 5 = ______

tens	ones

Regroup the tens. Draw the ones. Cross out cubes to show subtraction. Write the answer.

3. 21 − 7 = ______

tens	ones

4. 30 − 6 = ______

tens	ones

5. There are 42 icicles. 9 of them melt. How many icicles are left?

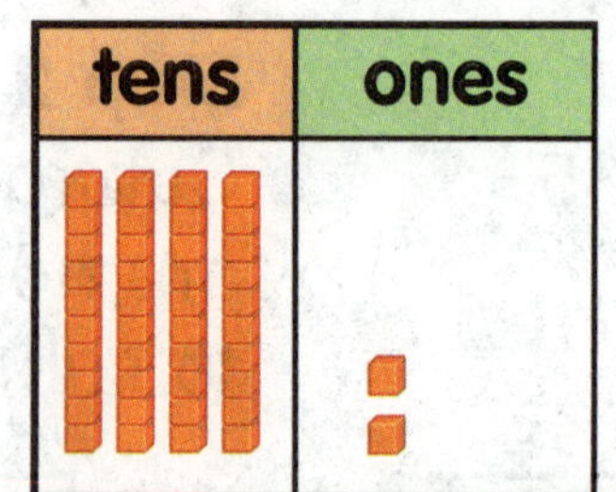

______ icicles

6. 31 people are at the skating rink. 8 people go home. How many people are left at the skating rink?

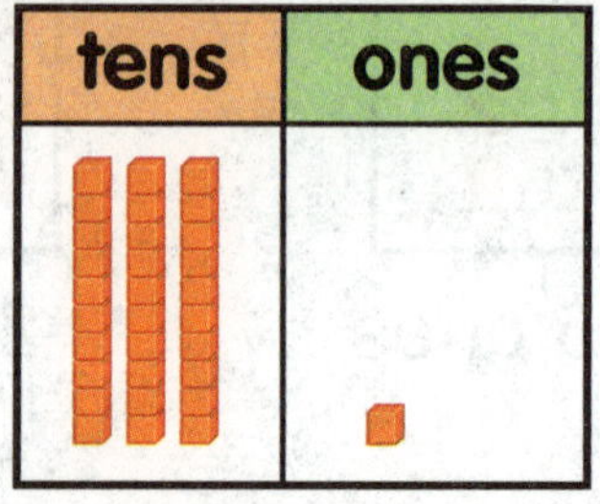

______ people

Test Practice

7. 23 − 6 = ______

tens	ones

29 ○ 17 ○ 13 ○ 10 ○

Math at Home Have your child use pennies to show you how to subtract 6 from 25.

Name ....

Subtract From a Two-Digit Number

Lesson 4

ESSENTIAL QUESTION
How can I subtract two-digit numbers?

Explore and Explain

tens	ones

_____ crabs

Teacher Directions: Jack saw 34 crabs on the beach. Use base-ten blocks to show that number. 8 crabs ran into the water. Regroup and take away that many blocks. Draw the blocks that are left. Write how many crabs were left on the beach.

See and Show

Helpful Hint
Cross out each number as you regroup it.

Mathematical PRACTICE

Find 34 − 6.

Step 1
Show 34.
Can you subtract 6 ones?

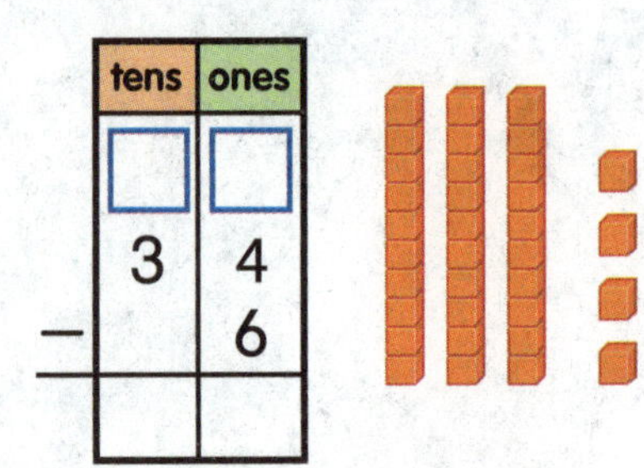

Step 2
Regroup 1 ten as 10 ones.
4 ones + 10 ones = 14 ones.

Step 3
Subtract the ones.

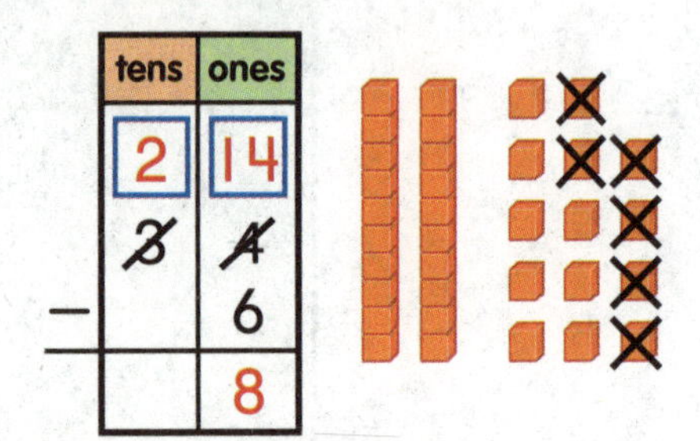

Step 4
Subtract the tens.

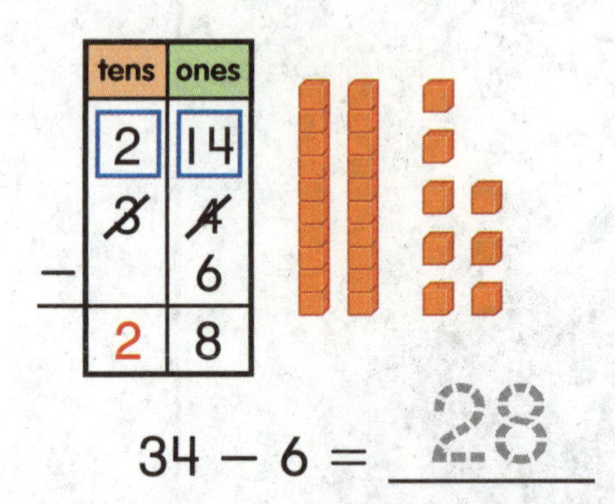

34 − 6 = 28

Use Work Mat 6 and base-ten blocks to subtract.

1.

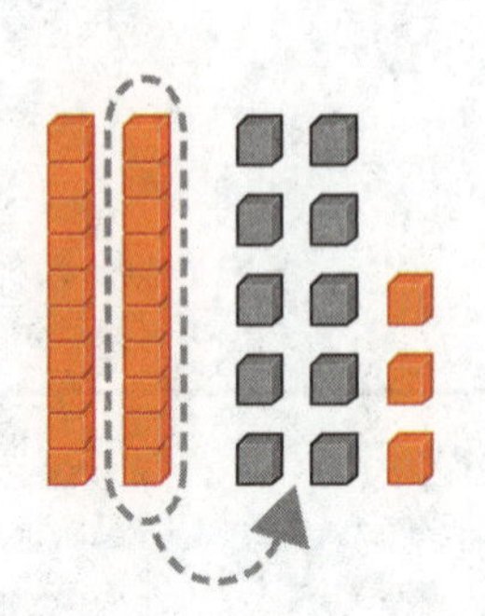

2.

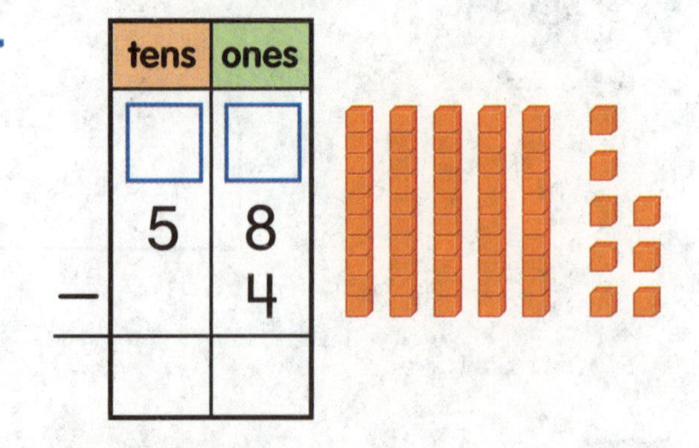

Talk Math How do you regroup 1 ten?

Name ....

CCSS

On My Own

Use Work Mat 6 and base-ten blocks to subtract.

3.

tens	ones
☐	☐
3	4
−	7

4.

tens	ones
☐	☐
7	3
−	3

5.

tens	ones
☐	☐
5	9
−	4

6. 88 − 2 = ____

7. 15 − 9 = ____

8. 31 − 6 = ____

9. 58 − 9 = ____

10. 66 − 7 = ____

11. 95 − 5 = ____

12. 14 − 7 = ____

13. 28 − 9 = ____

14. 34 − 6 = ____

Problem Solving

15. Tina puts 12 coins in a machine. The machine keeps 4 of the coins. How many coins does she get back?

_____ coins

16. There are 38 apples in the cafeteria. Students buy 9 apples for their lunches. How many apples are left?

_____ apples

17. A store has 76 shirts on display. The store sells 8 of the shirts. How many shirts are left?

_____ shirts

HOT Problem There are 23 boys and 33 girls at a beach. 7 boys and 4 girls leave the beach, how many boys and girls are left altogether? Explain.

____

____

____

____

Name ____________________

My Homework

Lesson 4
Subtract From a Two-Digit Number

Homework Helper

Need help? connectED.mcgraw-hill.com

Find 38 – 9.

Step 1 Can you subtract ones?
Step 2 Regroup 1 ten as 10 ones.
Step 3 Subtract the ones.
Step 4 Subtract the tens.

	tens	ones
	2	18
	~~3~~	~~8~~
–		9
	2	9

Practice

Subtract.

1.

	tens	ones
	☐	☐
	2	6
–		9

2.

	tens	ones
	☐	☐
	4	7
–		6

3.

	tens	ones
	☐	☐
	6	2
–		5

4.
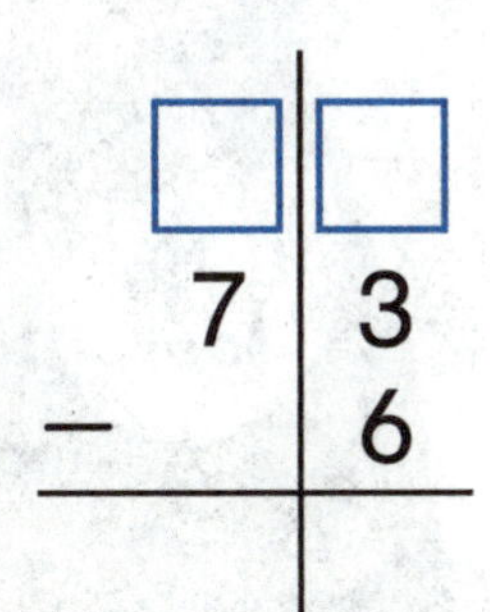

5.
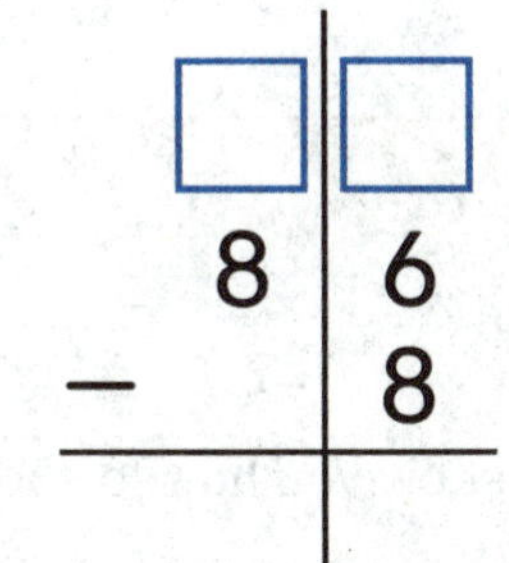

6.
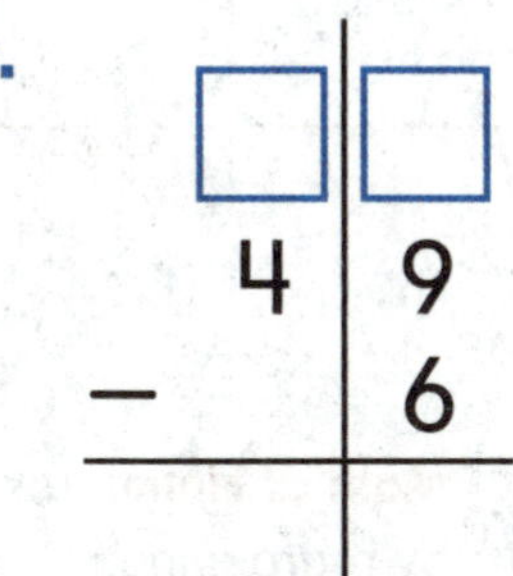

Subtract.

7. $\begin{array}{r} 44 \\ -\ 4 \\ \hline \end{array}$

8. $\begin{array}{r} 51 \\ -\ 8 \\ \hline \end{array}$

9. $\begin{array}{r} 22 \\ -\ 3 \\ \hline \end{array}$

10. There are 27 children playing outside. 7 children go inside. How many are left?

_______ children

11. There are 55 mice in the barn. A cat chases 9 of them away. How many mice are left?

_______ mice

Test Practice

12. Find the difference.

$43 - 9 =$ _______

52    44 ○   33    34 ○

I do too!

Math at Home Ask your child to show you how to subtract 8 from 27 by regrouping.

Name ______________________

Check My Progress

Vocabulary Check

Complete the sentence.

fact family   regroup

1. The numbers 3, 5, and 8 make up a __________.

Concept Check

Complete the fact family.

2. 19, 7, 12

___ + ___ = ___   ___ − ___ = ___

___ + ___ = ___   ___ − ___ = ___

Take apart a number to make a ten. Then subtract.

3. 83 − 52

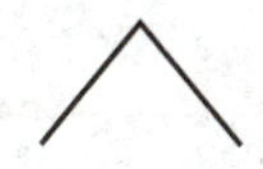

_____ _____

83 − _____ = _____

_____ − _____ = _____

So, 83 − 52 = _____.

4. 67 − 45

_____ _____

67 − _____ = _____

_____ − _____ = _____

So, 67 − 45 = _____.

Subtract.

5.

tens	ones
6	2
−	8

6.

tens	ones
4	3
−	7

7.

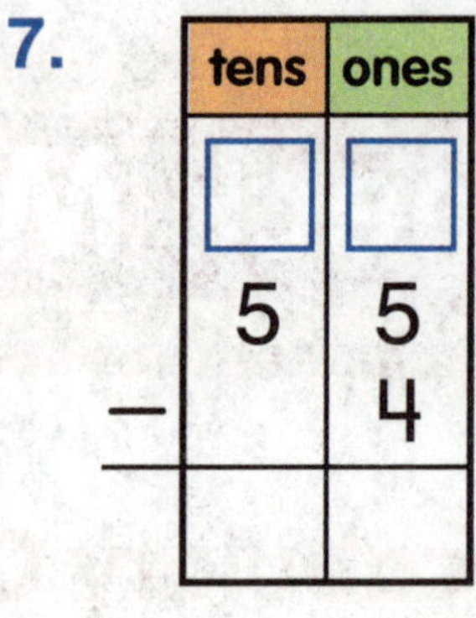

tens	ones
5	5
−	4

8.

$$\begin{array}{r|l} 2 & 5 \\ - & 6 \\ \hline & \end{array}$$

9.

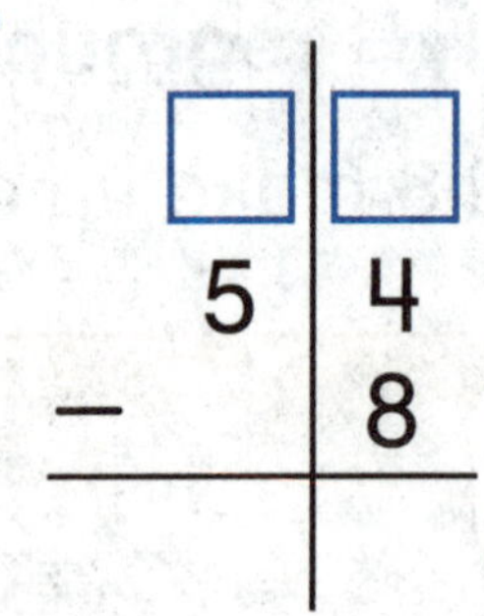

$$\begin{array}{r|l} 5 & 4 \\ - & 8 \\ \hline & \end{array}$$

10.

$$\begin{array}{r|l} 7 & 3 \\ - & 6 \\ \hline & \end{array}$$

11. There are 35 robins in a park. 14 of them fly away. How many robins are left in the park?

_____ robins

Test Practice

12. Yesterday, there were 77 tulips blooming in the garden. Today, 8 of them have wilted. How many tulips are still blooming?

69 ○ 78 ○ 79 ○ 85 ○

Name

Subtract Two-Digit Numbers

Lesson 5

ESSENTIAL QUESTION
How can I subtract two-digit numbers?

Explore and Explain

I can't see anything with these goggles.

tens	ones

_____ people

Teacher Directions: There are 42 people swimming at the pool. Use base-ten blocks to show those people. 13 people go home. Take away 13 base-ten blocks. Regroup if needed. Draw the base-ten blocks that are left in the chart. Write how many people are left. How many are left at the pool?

See and Show

Helpful Hint
Cross out numbers as you regroup them.

Find 52 − 17.

Step 1
Show 52.
Can you subtract 7 ones?

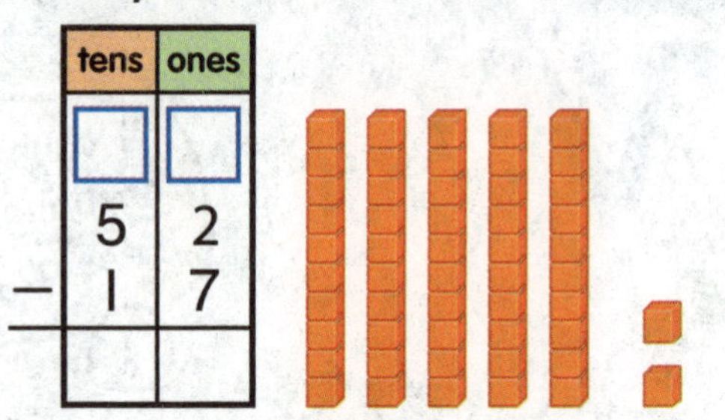

Step 2
Regroup 1 ten as 10 ones.
2 ones + 10 ones = 12 ones.

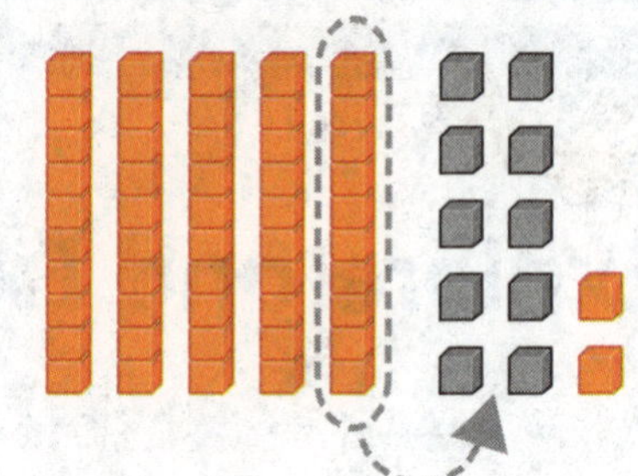

Step 3
Subtract the ones.

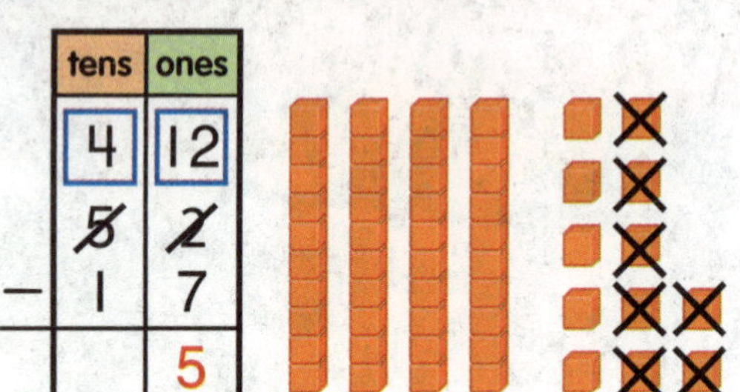

Step 4
Then subtract the tens.

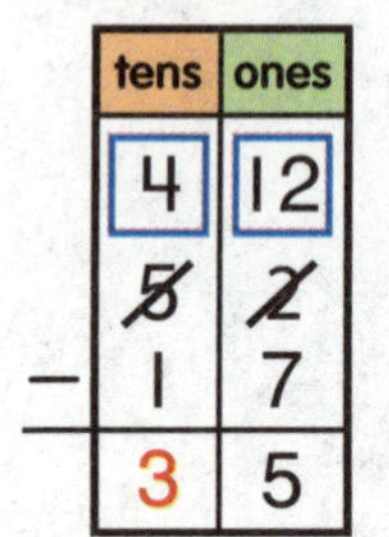

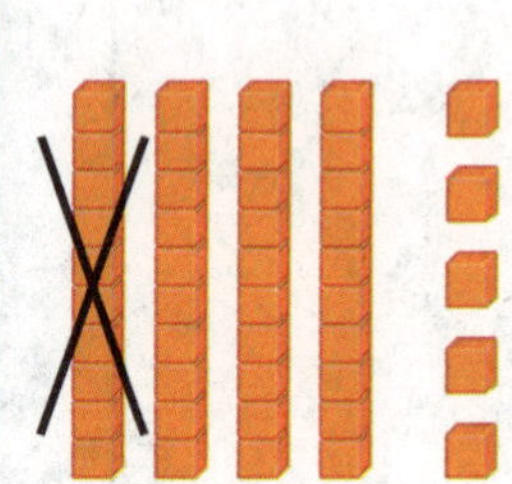

52 − 17 = 35

Use Work Mat 6 and base-ten blocks to subtract.

1.

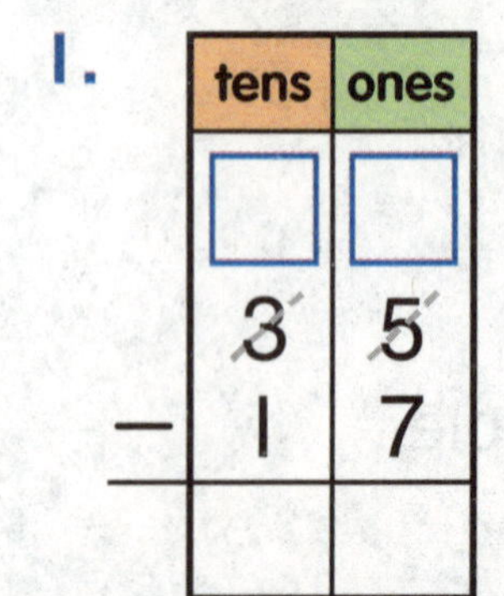

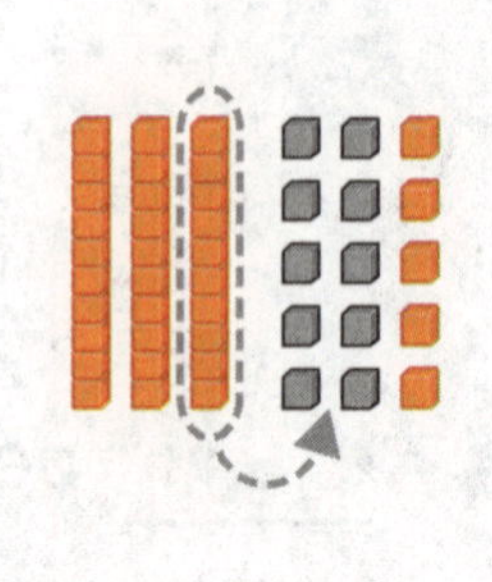

2.

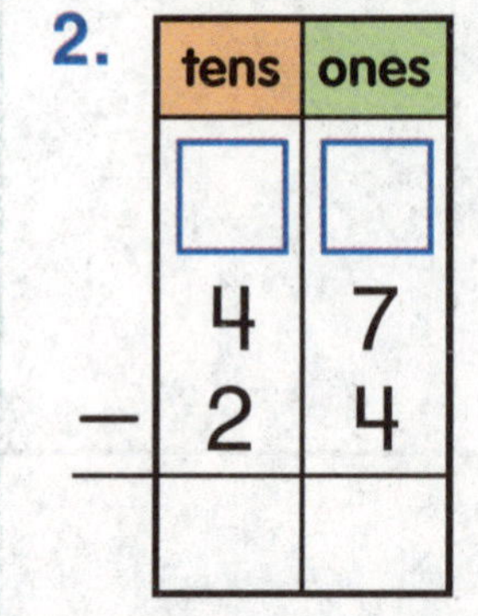

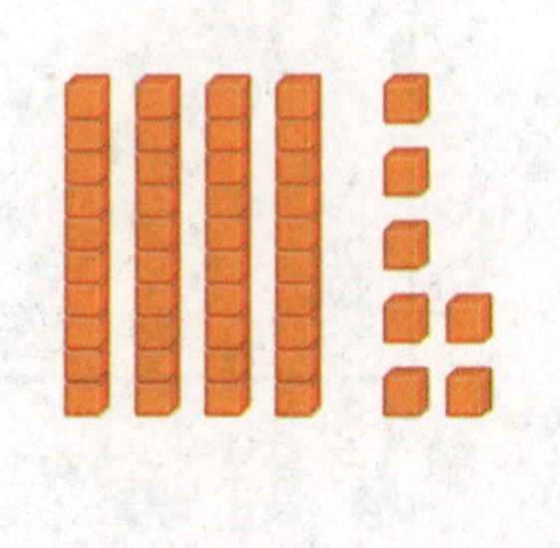

Talk Math How is subtracting 41 − 16 different than 41 − 6?

Name

On My Own

Use Work Mat 6 and base-ten blocks to subtract.

3.

tens	ones
2	6
− 1	9

4.

tens	ones
7	4
− 1	8

5.

tens	ones
4	6
− 2	3

6.
$$\begin{array}{r} 75 \\ -\ 16 \\ \hline \end{array}$$

7.
$$\begin{array}{r} 54 \\ -\ 28 \\ \hline \end{array}$$

8.
$$\begin{array}{r} 72 \\ -\ 35 \\ \hline \end{array}$$

9.
$$\begin{array}{r} 50 \\ -\ 24 \\ \hline \end{array}$$

10.
$$\begin{array}{r} 82 \\ -\ 37 \\ \hline \end{array}$$

11.
$$\begin{array}{r} 39 \\ -\ 29 \\ \hline \end{array}$$

12.
$$\begin{array}{r} 85 \\ -\ 46 \\ \hline \end{array}$$

13.
$$\begin{array}{r} 41 \\ -\ 15 \\ \hline \end{array}$$

14.
$$\begin{array}{r} 37 \\ -\ 11 \\ \hline \end{array}$$

Problem Solving

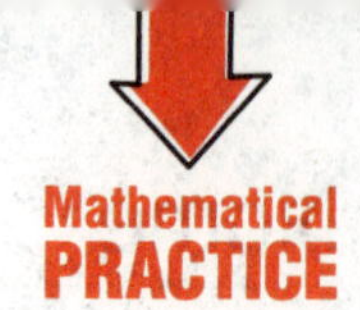

15. 35 people are in line for the diving board at the pool. 16 people jump off the diving board. How many people are still in line?

_____ people

16. Jason bought a pool pass for 45 trips to the pool. He has gone to the pool 28 times. How many more times can Jason go to the pool with his pass?

_____ times

HOT Problem Ella, May, and Sue each pour 24 cups of lemonade for their lemonade stand. They sell 23 of the cups of lemonade to their friends. How many cups are left to sell? Explain.

____

____

____

Name ....

My Homework

Lesson 5
Subtract Two-Digit Numbers

Homework Helper 

Need help? connectED.mcgraw-hill.com

Find 48 − 19.

Step 1 Can you subtract ones?

Step 2 Regroup 1 ten as 10 ones.

Step 3 Subtract the ones.

Step 4 Subtract the tens.

	tens	ones
	3	18
	~~4~~	~~8~~
−	1	9
	2	9

Practice

Subtract.

1.

	tens	ones
	□	□
	6	3
−	2	5

2.

	tens	ones
	□	□
	4	7
−	1	7

3.

	tens	ones
	□	□
	2	4
−	1	6

4.

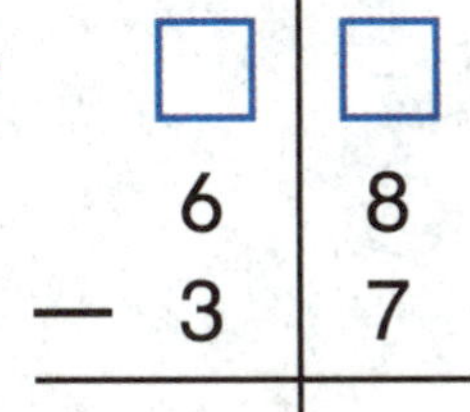

	□	□
	6	8
−	3	7

5.

	□	□
	5	3
−	2	4

6.

	□	□
	5	8
−	2	9

Subtract.

7. $\begin{array}{r} 54 \\ -\ 45 \\ \hline \end{array}$	8. $\begin{array}{r} 27 \\ -\ 9 \\ \hline \end{array}$	9. $\begin{array}{r} 63 \\ -\ 11 \\ \hline \end{array}$
10. $\begin{array}{r} 91 \\ -\ 59 \\ \hline \end{array}$	11. $\begin{array}{r} 35 \\ -\ 26 \\ \hline \end{array}$	12. $\begin{array}{r} 87 \\ -\ 42 \\ \hline \end{array}$

13. 83 people were at the water park. 29 people left. How many people were left at the water park?

Which way to the water?

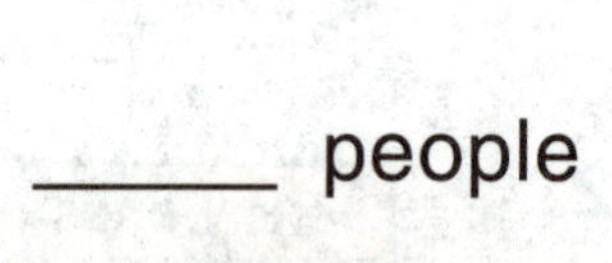

_____ people

Test Practice

14. 25 people brought dogs to the picnic. 18 of them were large dogs. How many small dogs were at the picnic?

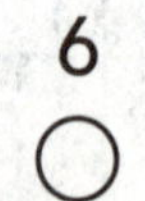

6	7	13	17
◯	◯	◯	◯

Math at Home Ask your child to show you how to subtract 24 from 41.

Name

Rewrite Two-Digit Subtraction

Lesson 6

ESSENTIAL QUESTION
How can I subtract two-digit numbers?

Explore and Explain

Spring has sprung!

54 − 36

	tens	ones
−		

_____ flowers

Teacher Directions: 54 flowers were planted in the garden. Write this number first. 36 flowers have bloomed. Write this number below. How many flowers still need to bloom? Rewrite the problem to subtract.

See and Show

You can rewrite a problem to subtract.
Find 83 – 28.

Step 1 Rewrite.

Step 2 Subtract.

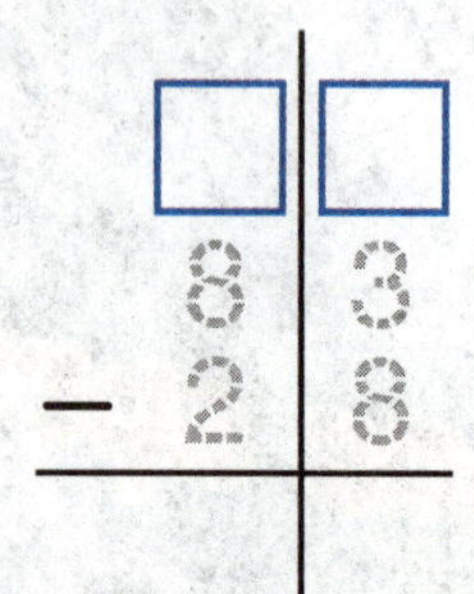

	7	13
	8	3
–	2	8
	5	5

Helpful Hint
Write the greater number on top. Write the other number below it.

Line up the ones and the tens.

Rewrite the problem. Subtract.

1. 38 – 19

2. 42 – 27

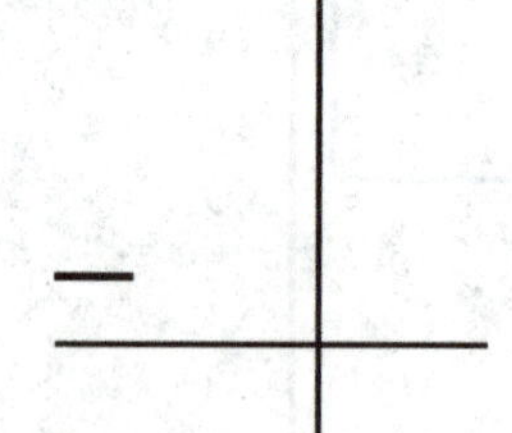

3. 53 – 38

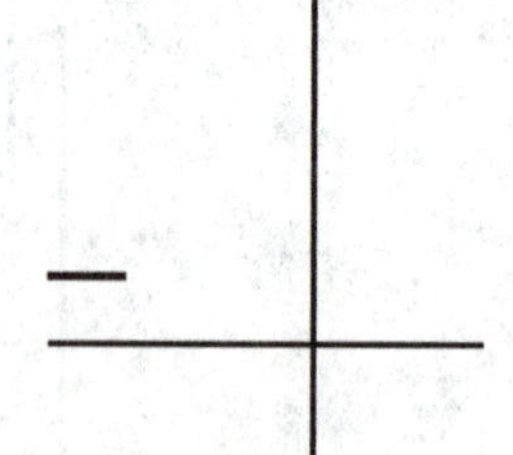

4. 82 – 46

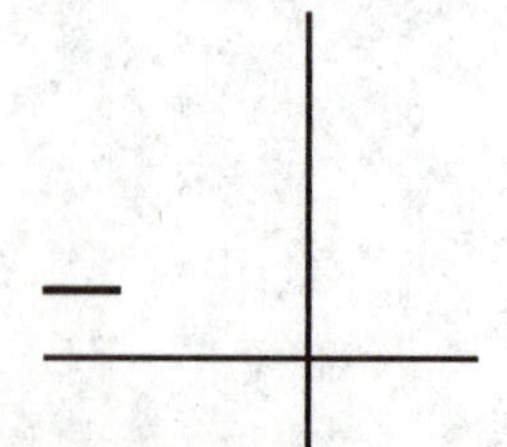

5. 74 – 38

6. 60 – 48

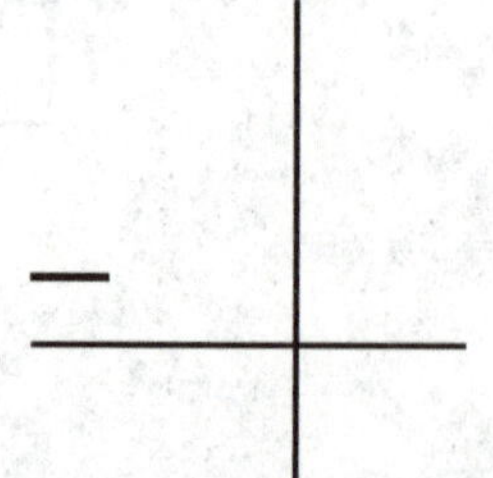

Talk Math Why is it helpful to rewrite subtraction?

Name

On My Own

Rewrite the problems. Subtract.

7. 62 – 38

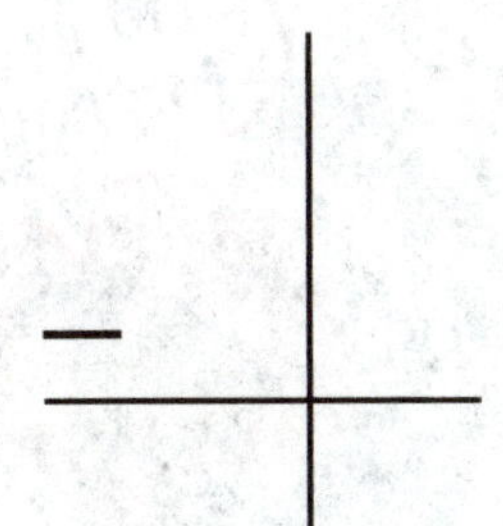

8. 74 – 39

9. 57 – 39

10. 88 – 49

11. 35 – 25

12. 62 – 38

13. 67 – 19

14. 83 – 39

15. 90 – 36

Problem Solving

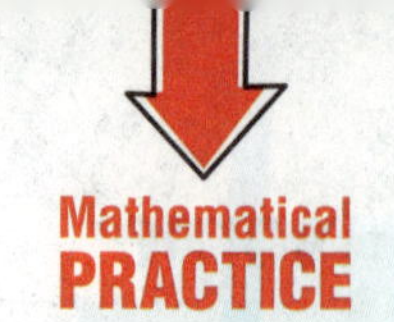

16. There are 64 flowers growing in Kyla's garden. She picks 18 flowers. How many flowers are left in the garden?

_____ flowers

17. Tyler kept track of the weather for 73 days. It was sunny 59 days. How many days were not sunny?

_____ days

18. Jackie has 32 houses on her street. She counted 17 that are brick. How many houses are not brick?

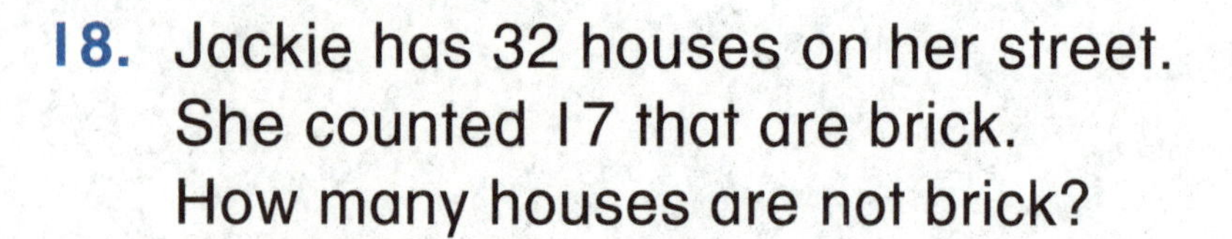

_____ houses

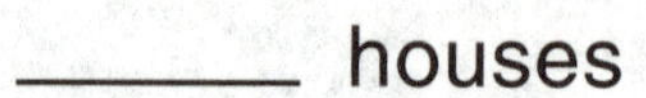

HOT Problem There are 91 children. 32 children play baseball. 21 children play soccer. How many children do not play baseball or soccer? Explain.

____

____

____

Name ....

My Homework

Lesson 6
Rewrite Two-Digit Subtraction

Homework Helper

Need help? connectED.mcgraw-hill.com

Rewrite to subtract 82 − 49.

7	12
8	2
− 4	9
3	3

Helpful Hint
When you subtract, always put the larger number on top.

Practice

Rewrite the problems. Subtract.

1. 74 − 25

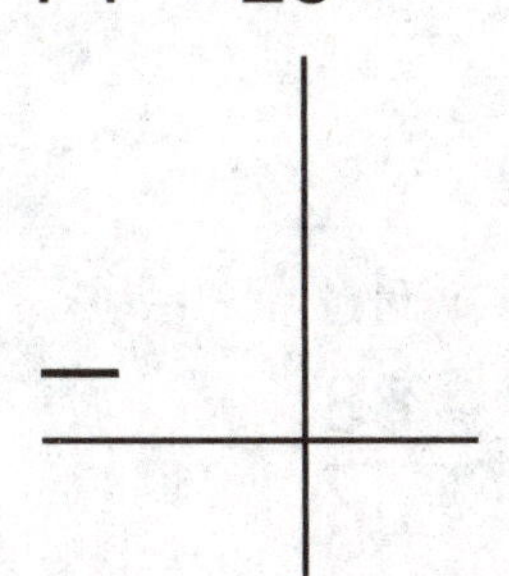

2. 60 − 37

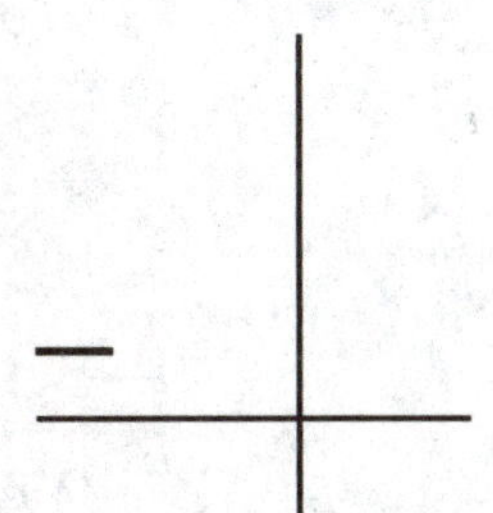

3. 86 − 48

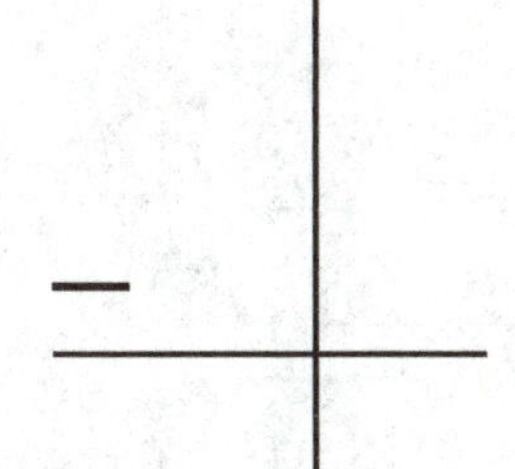

4. 45 − 28

−

5. 84 − 38

−

6. 37 − 18

−

Rewrite the problems. Subtract.

7. 72 − 37

−
______

8. 27 − 18

−
______

9. 68 − 39

−
______

10. 45 birds are flying. 17 birds land on the water. How many birds are still flying?

______ birds

Test Practice

11. Mark the answer that shows how to rewrite and solve 83 − 46.

7 13	7 13	3 16	3 16
83	83	46	46
−46	−46	−83	−83
36	37	37	43
○	○ 	○	○

Math at Home Together with your child think of a 2-digit subtraction problem in your lives. Have them solve it.

Name ....

Check Subtraction

Lesson 7

ESSENTIAL QUESTION
How can I subtract two-digit numbers?

Explore and Explain

Teacher Directions: There are 38 vegetables in the garden. Nancy picks 12 vegetables. How many vegetables are left? Show your work. Now check your work. Write the number left in the blue box. Write the number of vegetables picked in the yellow box. Add. What do you notice about the sum?

See and Show

Mathematical PRACTICE

You can check the answer to a subtraction problem. Check 25 − 10 = 15.

Helpful Hint
Add the number you subtracted to the difference. The sum should match the number you subtracted from.

Subtract		**Check by Adding**	
$\begin{array}{r} 25 \\ -\ 10 \\ \hline 15 \end{array}$	Add these numbers to check.	$\begin{array}{r} 15 \\ +\ 10 \\ \hline 25 \end{array}$	If this is the number you subtracted from, your answer is correct.

Subtract. Check by adding.

1. $\begin{array}{r} 36 \\ -\ 20 \\ \hline \end{array}$ $+$ ____

2. $\begin{array}{r} 50 \\ -\ 12 \\ \hline \end{array}$ $+$ ____

3. $\begin{array}{r} 21 \\ -\ 3 \\ \hline \end{array}$ $+$ ____

4. $\begin{array}{r} 32 \\ -\ 14 \\ \hline \end{array}$ $+$ ____

Talk Math Why does addition work as a check for subtraction?

Name

Helpful Hint
To check subtraction, add the number you subtracted and the difference.

CCSS

On My Own

Subtract. Check by adding.

5. 57 − 30 = ___ ; + ___

6. 60 − 10 = ___ ; + ___

7. 42 − 5 = ___ ; + ___

8. 75 − 41 = ___ ; + ___

9. 67 − 32 = ___ ; + ___

10. 40 − 26 = ___ ; + ___

11. 76 − 28 = ___ ; + ___

12. 56 − 28 = ___ ; + ___

Problem Solving

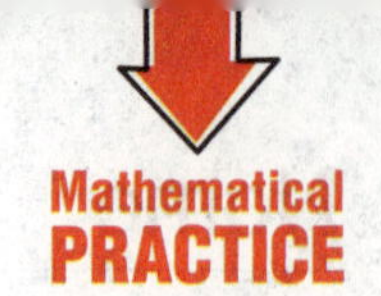

13. Ima went to the apple orchard with her class. The class picked 48 apples. They ate 17 of them. How many apples are left?

I was once a seed just like you!

_______ apples

14. The second grade classes are planting seeds. Roy's class plants 81 seeds. Jessica's class plants 66 seeds. How many more seeds does Roy's class plant than Jessica's class?

_______ seeds

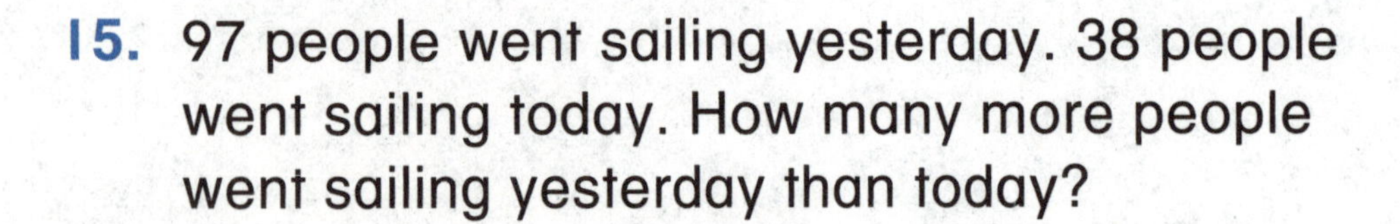

15. 97 people went sailing yesterday. 38 people went sailing today. How many more people went sailing yesterday than today?

_______ people

How do you check a subtraction problem?

____

____

____

Name ____________________

My Homework

Lesson 7
Check Subtraction

Homework Helper

Need help? connectED.mcgraw-hill.com

You can check the answer to a subtraction problem.

Check 37 − 15 = 22.

Subtract

$$\begin{array}{r} 37 \\ -\ 15 \\ \hline 22 \end{array}$$

Add these numbers to check.

Check by Adding

$$\begin{array}{r} 22 \\ +\ 15 \\ \hline 37 \end{array}$$

If this is the number you subtracted from, you are correct.

Practice

Subtract. Check by adding.

1. $\begin{array}{r} 67 \\ -\ 48 \\ \hline \end{array}$ + ____

2. $\begin{array}{r} 52 \\ -\ 36 \\ \hline \end{array}$ + ____

3. $\begin{array}{r} 80 \\ -\ 68 \\ \hline \end{array}$ + ____

4. $\begin{array}{r} 91 \\ -\ 45 \\ \hline \end{array}$ + ____

Solve. Check by adding.

5. There are 46 girls skating. There are 67 boys skating. How many more boys than girls are skating?

_______ boys

6. Molly checks out 21 books from the library. She returns 12. How many books does Molly still have?

_______ books

7. Reid kicked the football 20 times. He missed a field goal 6 times. How many times did he make a field goal?

_______ times

Test Practice

8. Mrs. Levine is 83 years old. Mrs. Smith is 67 years old. How many years older is Mrs. Levine than Mrs. Smith?

16 years	20 years	26 years	17 years
○	○	○	○

Math at Home Ask your child to find 43 − 16. Then have your child show you how to check the answer.

Name

Problem Solving

STRATEGY: Write a Number Sentence

Lesson 8

ESSENTIAL QUESTION
How can I subtract two-digit numbers?

There are 24 bees on a flower. 15 bees fly away. How many bees are now on the flower?

1 Understand

Underline what you know.
Circle what you need to find.

2 Plan

How will I solve the problem?

3 Solve

Write a number sentence.

24 ⊖ 15 = 9 bees

4 Check

Is my answer reasonable? Explain.

Practice the Strategy

84 children are at summer camp. 29 children leave to come home early. How many children are still at summer camp?

1 Understand Underline what you know. Circle what you need to find.

2 Plan How will I solve the problem?

3 Solve I will...

4 Check Is my answer reasonable? Explain.

Name

Apply the Strategy

Write a number sentence to solve.

1. There are 25 ants in an ant hill. 13 ants leave. How many ants are there now?

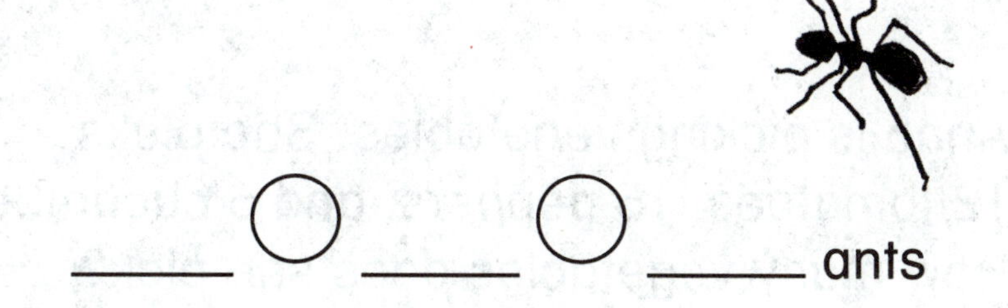

___ ◯ ___ ◯ ___ ants

2. There are 18 lions in the yard at the zoo. 7 lions run into the lion house. How many lions are left in the yard?

___ ◯ ___ ◯ ___ lions

3. Jay planted 12 daisies. His sister, Sarah, planted 29 daisies. How many more daisies did Sarah plant than Jay?

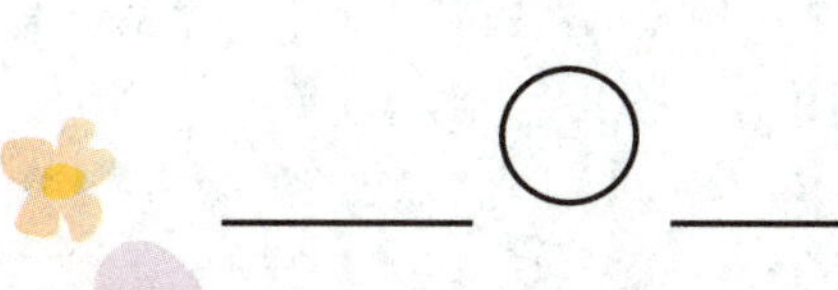

___ ◯ ___ ◯ ___ daisies

Review the Strategies

Choose a strategy

- Find a pattern.
- Write a number sentence.
- Act it out.

4. Brenda collects 39 leaves. Janet collects 45 leaves. How many more leaves did Janet collect than Brenda?

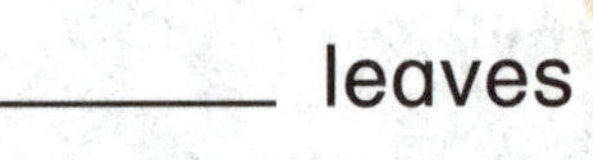

_______ leaves

5. Anna is picking vegetables. She picks 12 tomatoes, 15 peppers, and 5 cucumbers. How many vegetables does she pick?

_______ vegetables

6. Scott has 63 blueberries to put in a pie. He eats 21 of the blueberries. How many blueberries are in the pie?

_______ blueberries

7. Trey's dog likes dog treats. There are 60 treats in a box. He is allowed 2 treats a day. How many treats will he have left after 10 days?

_______ treats

Name

My Homework

Lesson 8
Problem Solving: Write a Number Sentence

Tara and her dad are baking bread for a bake sale. They make 34 loaves of bread. 11 loaves were sold. How many loaves are left?

1 Understand Underline what you know.
Circle what you need to find.

2 Plan How will I solve the problem?

3 Solve I will write a number sentence.

34 – 11 = 23 loaves

4 Check Is my answer reasonable?

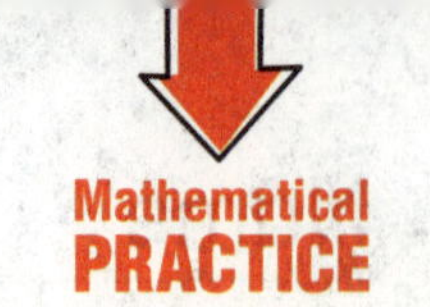

Practice

Write a number sentence to solve.

1. Emma has 31 ribbons for her hair. She loses 12 ribbons. How many does she have left?

   ____ ◯ ____ ◯ ____ ribbons

2. Mason's soccer team has 24 games this season. He has already played 18 games. How many games are left?

   ____ ◯ ____ ◯ ____ games

3. There are 18 ducks in the pond. 11 ducks get out of the pond. How many are still in the pond?

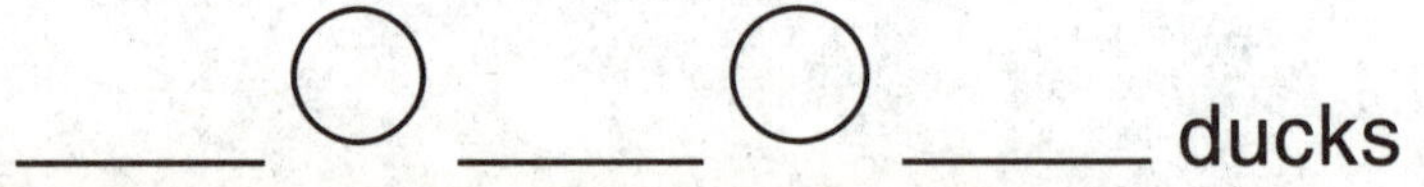

   ____ ◯ ____ ◯ ____ ducks

Test Practice

4. Karen took a walk on the beach and counted 25 sandcastles. On the way back she counted again but some were gone. There were only 16 left. How many had disappeared while she walked?

   ◯ 11 ◯ 9 ◯ 19 ◯ 41

Math at Home Ask your child to subtract 52 − 17. Then have your child show you how to check the answer.

Name

Two-Step Word Problems

Lesson 9

ESSENTIAL QUESTION
How can I subtract two-digit numbers?

Explore and Explain

Wipe out!

______ people

Teacher Directions: Use base-ten blocks to solve the problem. Draw the blocks to show your work. There are 62 people sledding. 13 people leave to go home. 12 people go inside for hot cocoa. How many people are still sledding?

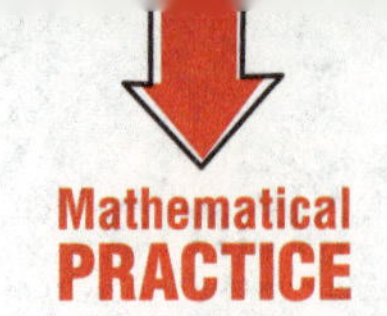

See and Show

Some word problems take two-steps to solve.
There are 87 books at the library. 23 of the books are checked out. 4 of those books are returned. How many books are at the library now?

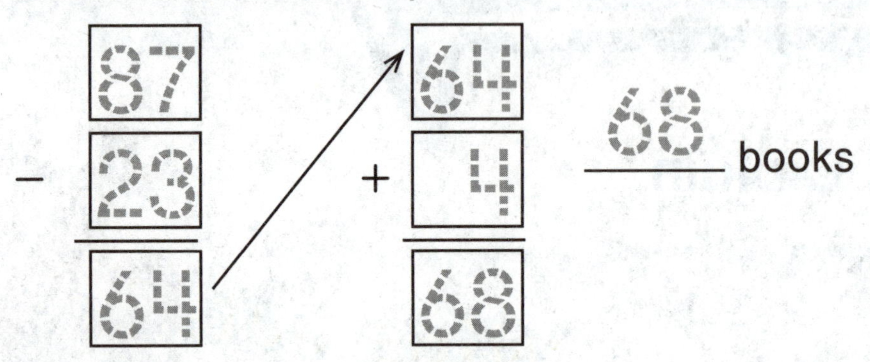

Solve each two-step word problem.

I read books every day. I eat a few too!

1. Hailey has 33 books. She gives 14 books to her cousin. She gives 5 books to her sister. How many books does Hailey have left?

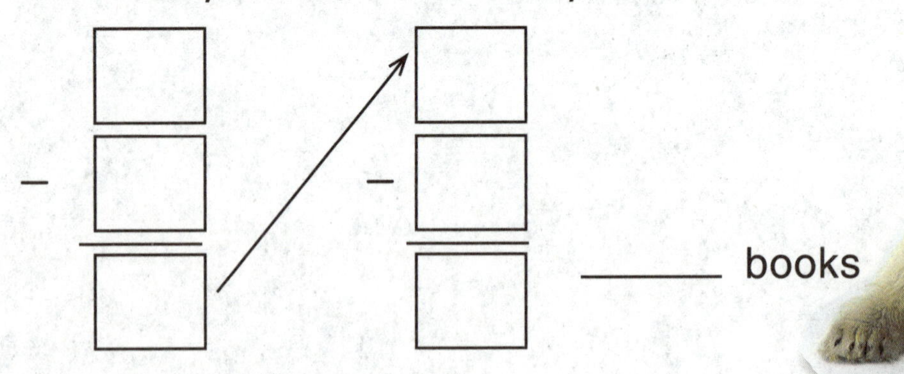

2. Dale invites 25 people to his party. 8 people cannot come. 5 people come that were not invited. How many people come to the party?

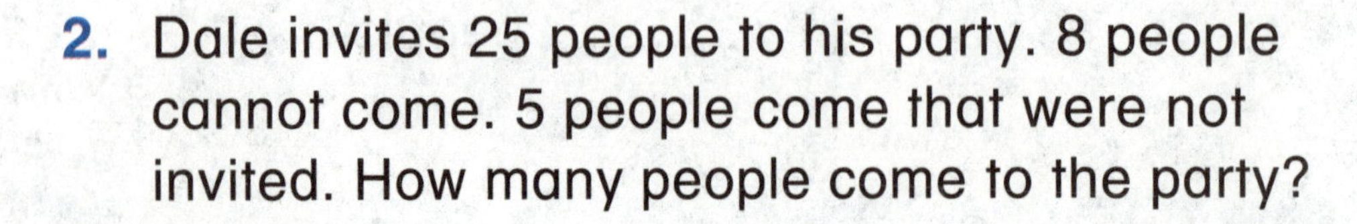

_____ people

Talk Math How do you solve a two-step problem?

Name

On My Own

Solve each two-step word problem.

3. Maryanne builds a tower out of blocks. She uses 89 blocks. 18 blocks fall off the tower. Her brother knocks 14 more blocks off the tower. How many blocks are still on the tower?

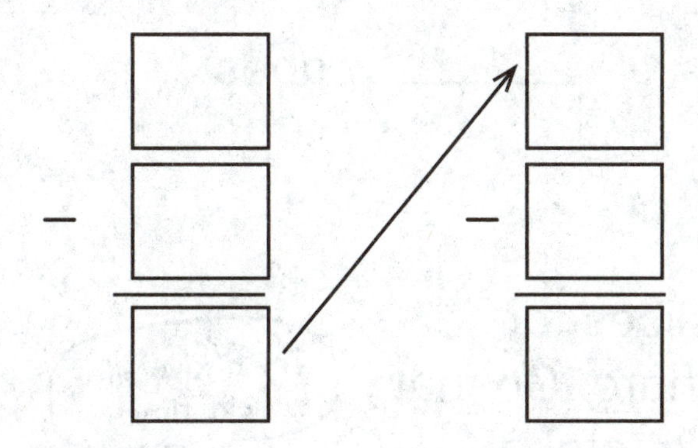

_____ blocks

4. Ray has 56 comics. He gives 12 to his friend. He borrows 14 from his little brother. How many comics does Ray have now?

_____ comics

5. There are 39 penguins on the iceberg. 11 penguins jump into the water. 4 more penguins jump onto another iceberg. How many penguins are left?

_____ penguins

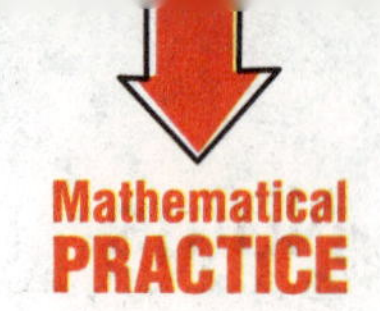

Real World Problem Solving

Solve each two-step word problem.

6. There are 48 grapes in the bowl. Mia eats 18 grapes. Her mom adds 22 grapes to the bowl. How many grapes are in the bowl now?

_____ grapes

7. Connor's mom cut 32 apple slices for his two brothers and him to share. Connor ate 5 slices. His brothers ate a total of 18 slices. How many apple slices were left?

_____ slices

HOT Problem Write a two-step word problem using the number sentences $13 + 10 = 23$ and $23 - 5 = 18$.

____

____

____

Name ....

My Homework

Lesson 9
Two-Step Word Problems

Homework Helper

Need help? connectED.mcgraw-hill.com

Katie's Coffee Shop sells 84 cups of cocoa on Saturday. 23 cups are sold in the morning. 35 cups are sold in the afternoon. How many cups are sold in the evening?

$$\begin{array}{r} 84 \\ -\ 23 \\ \hline 61 \end{array} \qquad \begin{array}{r} 61 \\ -\ 35 \\ \hline 26 \end{array}$$

Helpful Hint
Some problems need to be worked in two steps.

Practice

Solve each two-step word problem.

1. 56 dogs are at the dog park. 14 dogs go home. 15 dogs come to the park. How many dogs are at the dog park now?

   _______ dogs

2. The animal shelter has 33 cats. 9 cats are adopted on Friday. 12 cats are adopted on Saturday. How many cats are left at the animal shelter?

   _______ cats

Solve each two-step word problem.

3. Mr. Henry's class has 24 students.
3 students are absent today.
4 students are late to school.
How many students were on time?

_______ students

4. Abigail has 42 erasers. She gives 18 to Avery. She gives 15 to Mackenzie. How many erasers does Abigail have left?

_______ erasers

5. 15 children got on the bus. 6 children got off of the bus. 8 more children got on the bus. How many children are on the bus now?

_______ children

Test Practice

6. Caleb had 89 race cars. He lost 23 of them. He gave 14 to his friend. How many race cars does Caleb have left?

75	66	52	51
○	○	○	

Math at Home Create a two-step word problem for your child to solve. Have him or her show you how to solve it.

Name

Fluency Practice

Subtract.

1. 16 − 3	**2.** 26 − 6	**3.** 13 − 8
4. 15 − 9	**5.** 12 − 5	**6.** 19 − 5
7. 11 − 9	**8.** 12 − 6	**9.** 14 − 8
10. 19 − 8	**11.** 18 − 9	**12.** 17 − 3
13. 19 − 1	**14.** 15 − 5	**15.** 17 − 4

Fluency Practice

Subtract.

1. $\begin{array}{r} 13 \\ -\ 6 \\ \hline \end{array}$

2. $\begin{array}{r} 13 \\ -\ 7 \\ \hline \end{array}$

3. $\begin{array}{r} 18 \\ -\ 8 \\ \hline \end{array}$

4. $\begin{array}{r} 15 \\ -\ 6 \\ \hline \end{array}$

5. $\begin{array}{r} 17 \\ -\ 5 \\ \hline \end{array}$

6. $\begin{array}{r} 14 \\ -\ 8 \\ \hline \end{array}$

7. $\begin{array}{r} 20 \\ -\ 9 \\ \hline \end{array}$

8. $\begin{array}{r} 16 \\ -\ 7 \\ \hline \end{array}$

9. $\begin{array}{r} 11 \\ -\ 10 \\ \hline \end{array}$

10. $\begin{array}{r} 12 \\ -\ 9 \\ \hline \end{array}$

11. $\begin{array}{r} 18 \\ -\ 7 \\ \hline \end{array}$

12. $\begin{array}{r} 12 \\ -\ 7 \\ \hline \end{array}$

13. $\begin{array}{r} 16 \\ -\ 2 \\ \hline \end{array}$

14. $\begin{array}{r} 13 \\ -\ 3 \\ \hline \end{array}$

15. $\begin{array}{r} 19 \\ -\ 7 \\ \hline \end{array}$

Name ______________________

My Review

Vocabulary Check

Complete each sentence. Use words from the box.

subtract	fact family	regroup	difference

1. A ____________________ is a group of 3 numbers that work together.

2. You can ____________________ a ten into ten ones.

3. When you ____________________, you take something away from another number.

4. The answer you get from subtracting is called the ____________________.

Write an example of a fact family.

5.

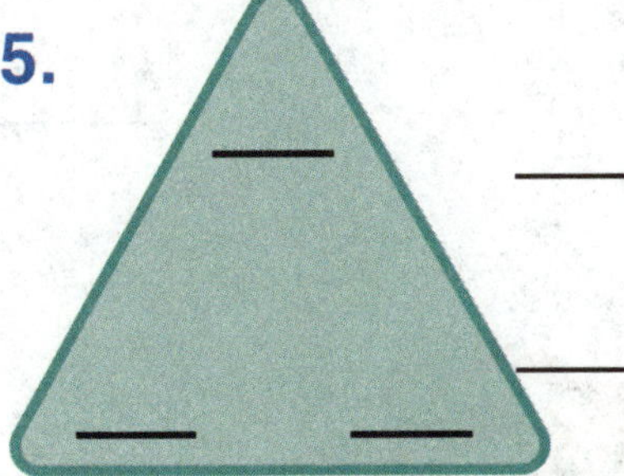

____ + ____ = ____   ____ − ____ = ____

____ + ____ = ____   ____ − ____ = ____

Concept Check

Subtract.

6. $\begin{array}{r} 27 \\ -18 \\ \hline \end{array}$

7. $\begin{array}{r} 32 \\ -17 \\ \hline \end{array}$

8. $\begin{array}{r} 26 \\ -19 \\ \hline \end{array}$

9. $\begin{array}{r} 48 \\ -28 \\ \hline \end{array}$

10. $\begin{array}{r} 83 \\ -37 \\ \hline \end{array}$

11. $\begin{array}{r} 80 \\ -26 \\ \hline \end{array}$

12. $\begin{array}{r} 74 \\ -25 \\ \hline \end{array}$

13. $\begin{array}{r} 71 \\ -37 \\ \hline \end{array}$

Rewrite the numbers. Subtract.

14. 48 − 29

15. 26 − 17

16. 53 − 37

Subtract. Check by adding.

17. $\begin{array}{r} 73 \\ -27 \\ \hline \end{array}$ $+$ ____

18. $\begin{array}{r} 63 \\ -36 \\ \hline \end{array}$ $+$ ____

Hit the beach!

Name ....

Problem Solving

19. Max has 34 marbles. He gives 19 marbles away. How many marbles does Max have left?

_______ marbles

20. There are 18 alligators around a pond. 10 alligators slide into the water. 2 come back out. How many alligators are around the pond now?

_______ alligators

21. Jill collects 68 golf balls. She gives 39 of them to her grandpa. She loses 8 of them. How many golf balls does Jill have left?

_______ golf balls

Test Practice

22. Matt sees 25 planes at the air show. 6 of the planes are red. 8 are white. The rest are blue. How many planes are blue?

39	19	14	11
○	○	○	○

Reflect

Chapter 4

Answering the Essential Question

Show the ways to subtract.
Check your work.

Subtract from a two-digit number. $\begin{array}{r} 25 \\ -\ 4 \\ \hline \end{array}$	Subtract with regrouping. $\begin{array}{r} 32 \\ -\ 6 \\ \hline \end{array}$
Subtract without regrouping. $\begin{array}{r} 48 \\ -\ 35 \\ \hline \end{array}$	Subtract with regrouping. $\begin{array}{r} 71 \\ -\ 19 \\ \hline \end{array}$

ESSENTIAL QUESTION

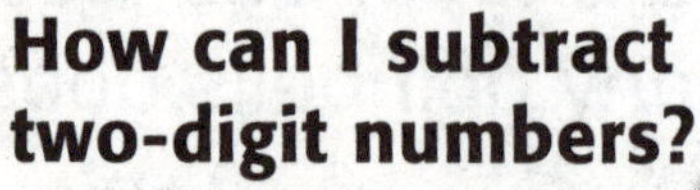
How can I subtract two-digit numbers?

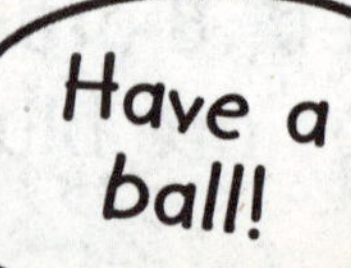

Let's get rolling!

Chapter

Place Value to 1,000

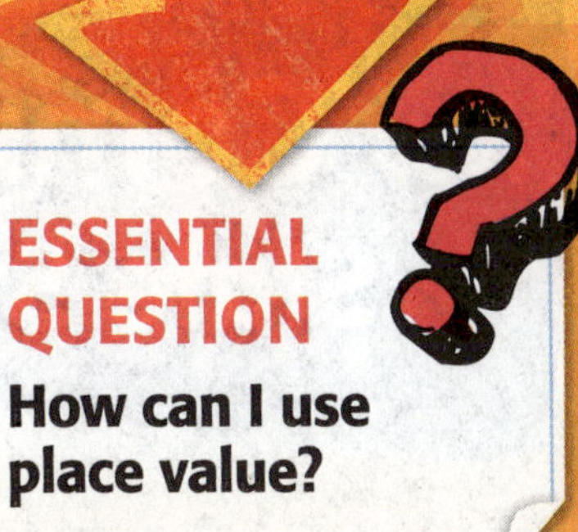

My Common Core State Standards

Number and Operations in Base Ten

CCSS

2.NBT.1 Understand that the three digits of a three-digit number represent amounts of hundreds, tens, and ones; e.g., 706 equals 7 hundreds, 0 tens, and 6 ones. Understand the following as special cases:

2.NBT.1a 100 can be thought of as a bundle of ten tens — called a "hundred."

2.NBT.1b The numbers 100, 200, 300, 400, 500, 600, 700, 800, 900 refer to one, two, three, four, five, six, seven, eight, or nine hundreds (and 0 tens and 0 ones).

2.NBT.2 Count within 1000; skip count by 5s, 10s, and 100.

2.NBT.3 Read and write numbers to 1000 using base-ten numerals, number names, and expanded form.

2.NBT.4 Compare two three-digit numbers based on meanings of the hundreds, tens, and ones digits, using >, =, and < symbols to record the results of comparisons.

2.NBT.8 Mentally add 10 or 100 to a given number 100–900, and mentally subtract 10 or 100 from a given number 100–900.

Standards for Mathematical PRACTICE

1. Make sense of problems and persevere in solving them.
2. Reason abstractly and quantitatively.
3. Construct viable arguments and critique the reasoning of others.
4. Model with mathematics.
5. Use appropriate tools strategically.
6. Attend to precision.
7. Look for and make use of structure.
8. Look for and express regularity in repeated reasoning.

= focused on in this chapter

Name ______________________________

Am I Ready?

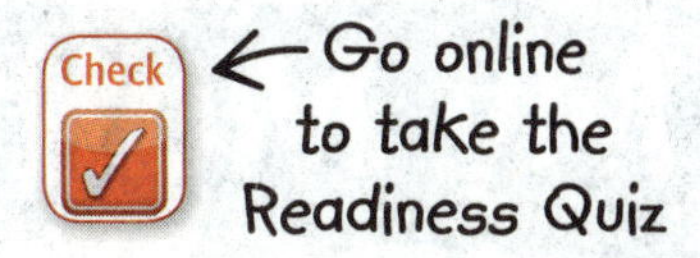

Write the number.

1. 

______

2. 

______

Compare. Write >, <, or =.

3. 27 ◯ 36

4. 15 ◯ 15

5. 45 ◯ 29

6. 12 ◯ 20

7. Tia counts the shoes in her closet. Each pair has 2 shoes. Show how she counts.

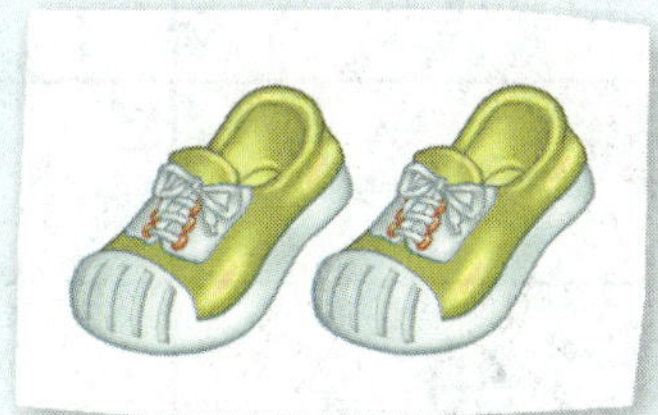

______, 4, ______, ______, 10

How Did I Do? Shade the boxes to show the problems you answered correctly.

1	2	3	4	5	6	7

Name

My Math Words

Review Vocabulary

ones tens

Mark an X to show the value of the underlined digit. Then write the value of the digit.

	Tens	Ones	Value
57		X	7
89			
20			
32			
41			
74			
68			

Write a sentence about a review word.

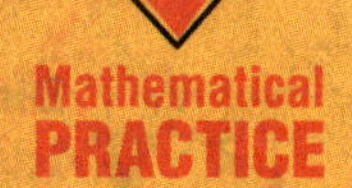

My Vocabulary Cards

Lesson 5-7

compare

3 = 3

Lesson 5-3

digit

2 digits

Lesson 5-7

equal to (=)

400 = 400

Lesson 5-3

expanded form

536 = 500 + 30 + 6

Lesson 5-7

greater than (>)

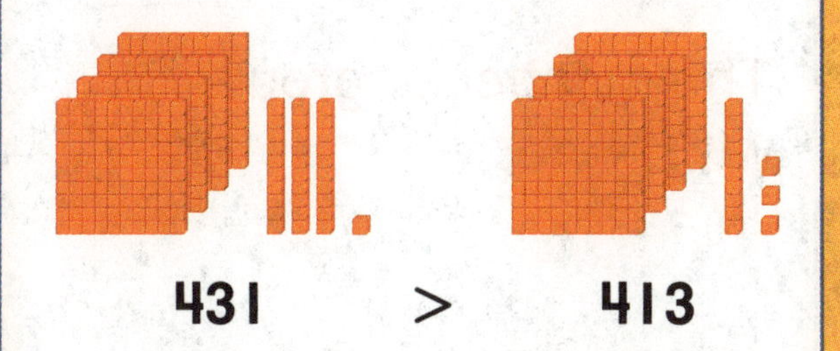

431 > 413

Lesson 5-1

hundreds

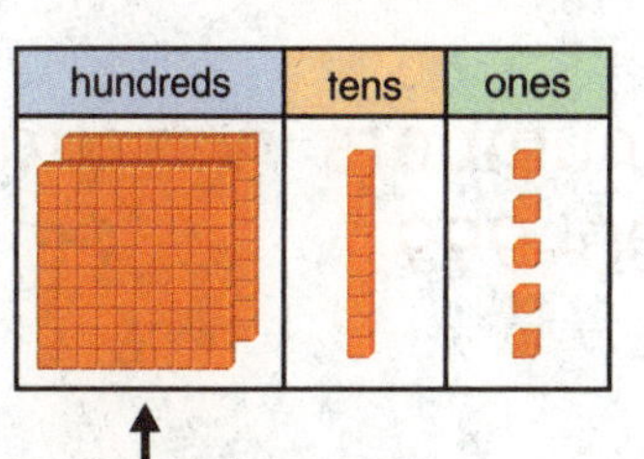

2 hundreds

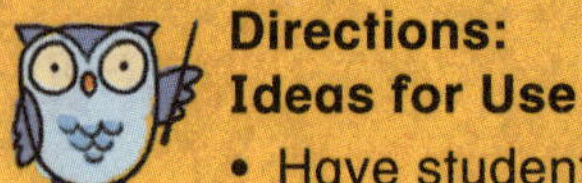

Directions:
Ideas for Use

- Have students group 2 or 3 common words. Ask them to add a word that is unrelated to the group. Have them work with a friend to name the unrelated word.
- Ask students to arrange cards to show words that are opposites. Have them explain the meaning of their opposite words.

A symbol used to write numbers. The ten digits are 0, 1, 2, 3, 4, 5, 6, 7, 8, and 9.

Look at numbers to see how they are alike or different.

The representation of a number as a sum that shows the value of each digit.

Have the same value as or is the same value as.

The numbers in the range of 100–999.

The number or group with more.

My Vocabulary Cards

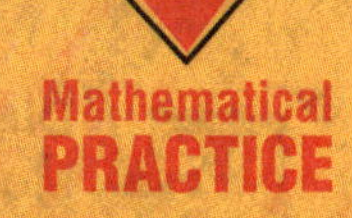

Mathematical PRACTICE

Lesson 5-7

less than (<)

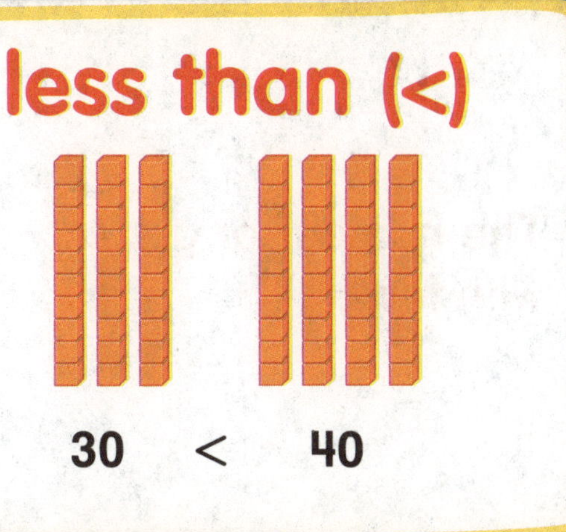

30 < 40

Lesson 5-1

ones

2 tens and 3 ones

20 + 3

Lesson 5-3

place value

hundreds	tens	ones
7	8	5

700 + 80 + 5 =
785

Lesson 5-1

tens

2 tens and 3 ones

20 + 3

Lesson 5-5

thousand

thousands	hundreds	tens	ones

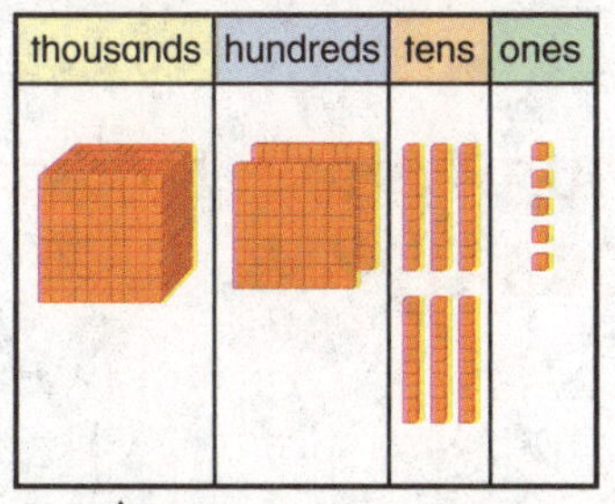

1 thousand

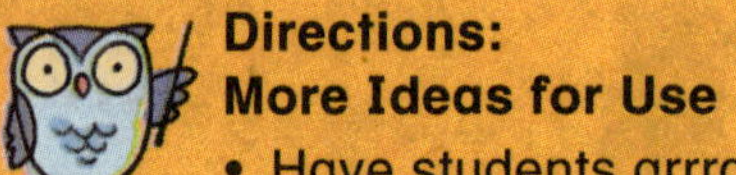

Directions:
More Ideas for Use

- Have students arrrange the cards in alphabetical order.
- Ask students to use the blank card to write their own vocabulary card.

The numbers in the range of 0–9. It is the place value of a number.

The number or group with fewer.

The numbers in the range of 10–99. It is the place value of a number.

The value given to a digit by its place in a number.

The numbers in the range of 1,000–1,999.

My Foldable

FOLDABLES® Follow the steps on the back to make your Foldable.

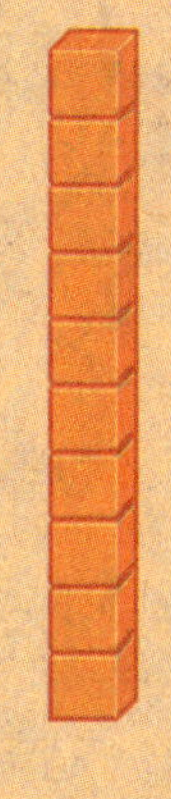

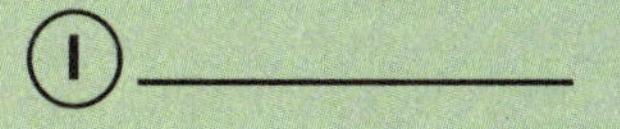

③ ______________

④ ______________

1

2

3

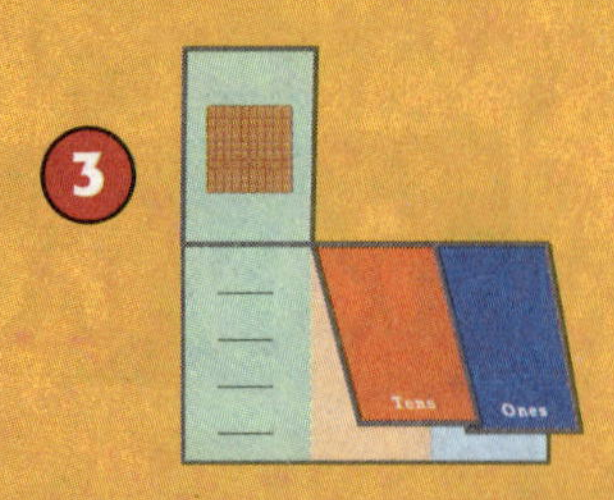

Ones

Tens

Hundreds

Name

Hundreds

Lesson 1

ESSENTIAL QUESTION
How can I use place value?

Explore and Explain

Watch

Tools

Teacher Directions: Color each stamp yellow and count the stamps by ones. Circle groups of ten in green to show tens. Count by tens to 100 with a classmate. Circle one hundred stamps in red to show a hundred.

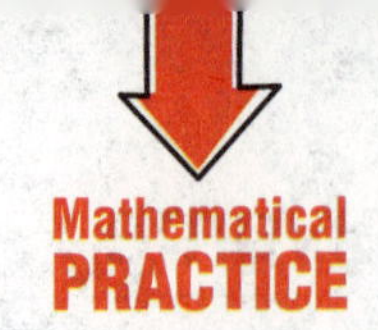

See and Show

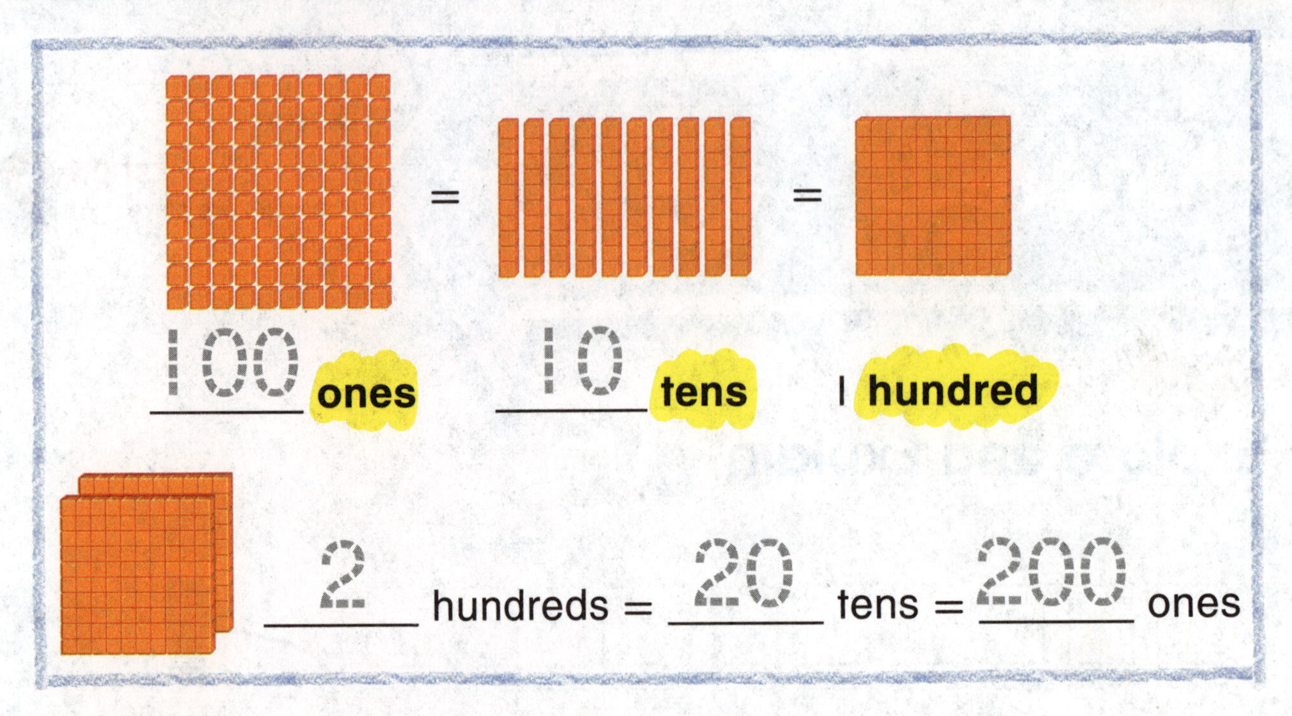

Write how many hundreds, tens, and ones.

1. _____ hundred = _____ tens = _____ ones

2. _____ hundreds = _____ tens = _____ ones

Write how many tens and hundreds.

3. 500 ones = _____ tens = _____ hundreds

4. 600 ones = _____ tens = _____ hundreds

Is 100 the same as ten tens?
How do you know?

Name ______________________

CCSS

On My Own

Write how many hundreds, tens, and ones.

5. 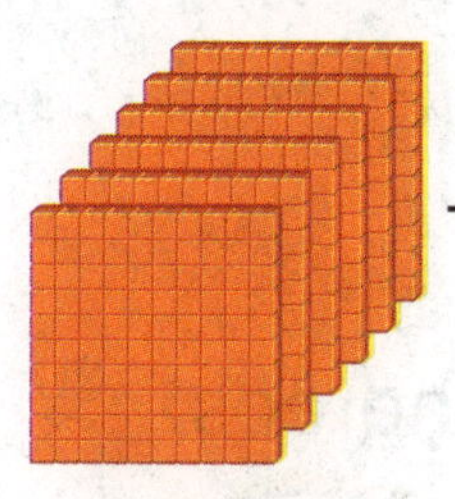 _______ hundreds = _______ tens = _______ ones

6. 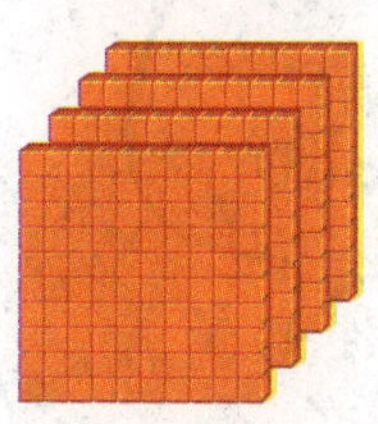 _______ hundreds = _______ tens = _______ ones

7. 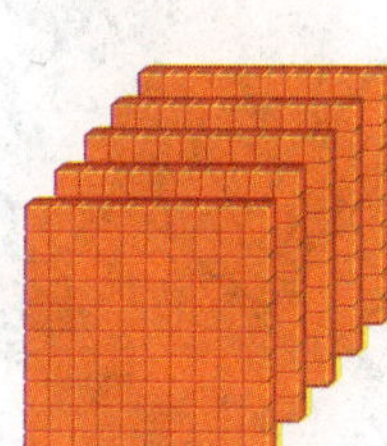 _______ hundreds = _______ tens = _______ ones

Write how many tens and hundreds.

8. 800 ones = _______ tens = _______ hundreds

9. 700 ones = _______ tens = _______ hundreds

Write how many hundreds, tens, or ones.

10. 500 = _______ tens

11. 200 = _______ ones

12. 400 = _______ hundreds

13. 300 = _______ tens

Problem Solving

14. Sara has a collection of coins. She puts them in 30 piles of 10. How many hundreds does she have?

_______ hundreds

15. Kenny places his rock collection in 20 groups of 10. How many ones does he have?

_______ ones

16. Andreas counts 300 leaves on a tree. How many tens does he count?

_______ tens

How many ones are in 4 hundreds? How do you know?

____

____

____

Name ______________________

My Homework

Lesson 1
Hundreds

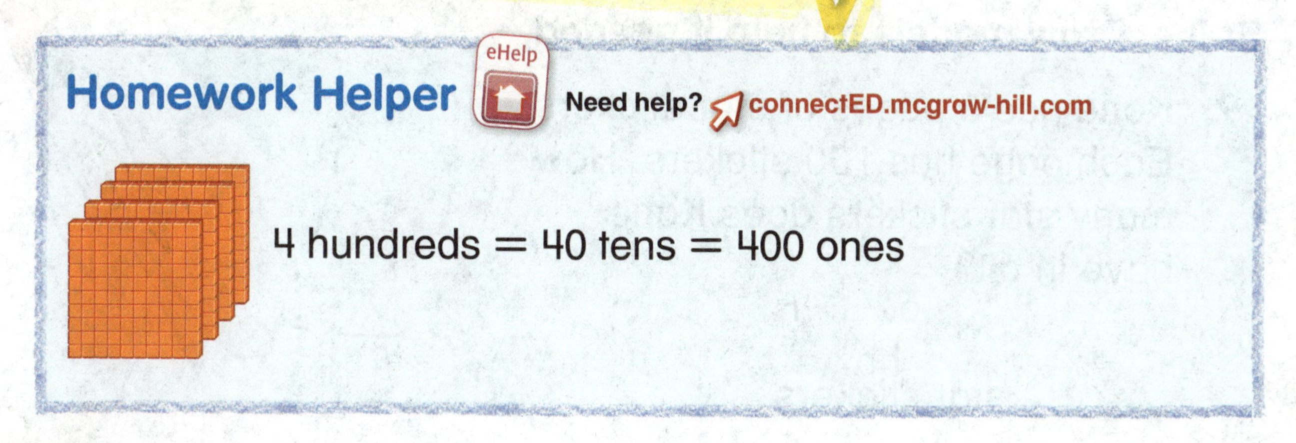

Homework Helper

eHelp Need help? connectED.mcgraw-hill.com

4 hundreds = 40 tens = 400 ones

Practice

Write how many hundreds, tens, and ones.

1. ______ hundreds = ______ tens = ______ ones

2. ______ hundreds = ______ tens = ______ ones

3. ______ hundreds = ______ tens = ______ ones

Write how many tens and hundreds.

4. 900 ones = ______ tens = ______ hundreds

Write how many hundreds, tens, or ones.

5. 800 = _______ tens

6. 400 = _______ ones

7. 700 = _______ hundreds

8. 600 = _______ tens

Solve. Draw models to help if needed.

9. Katie has 5 pages of star stickers. Each page has 100 stickers. How many star stickers does Katie have in all?

_______ star stickers

10. Avante has 7 baskets of 100 beads. How many hundreds of beads does Avante have?

_______ hundreds

Vocabulary Check

11. Circle the digit that shows **hundreds**. Underline the **ones** digit once. Underline the **tens** digit twice.

600

Math at Home Ask your child to count by hundreds to 1,000.

Name ....

Hundreds, Tens, and Ones

Lesson 2

ESSENTIAL QUESTION
How can I use place value?

Explore and Explain

Tools

hundreds	tens	ones
1	2	3

123 cans

Teacher Directions: Mya collected cans to recycle. The amount she collected had 3 hundreds, 6 tens, and 7 ones. Use base-ten blocks to show the number in the chart. Write the number.

Mathematical PRACTICE

See and Show

There are 427 pennies in this jar. Use hundreds, tens, and ones to show 427.

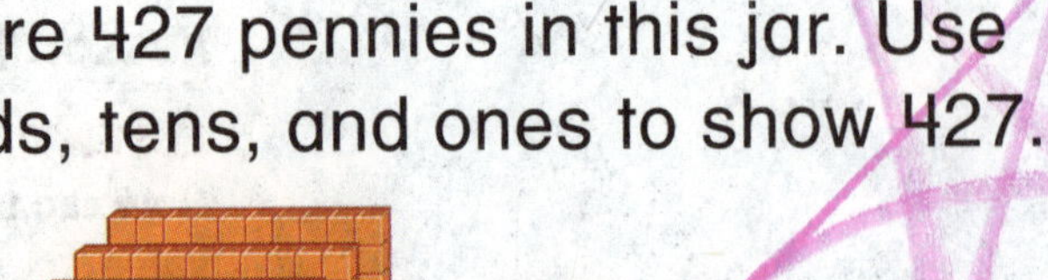

hundreds	tens	ones
4	2	7

427 four hundred twenty-seven

Write how many hundreds, tens, and ones. Then write the number.

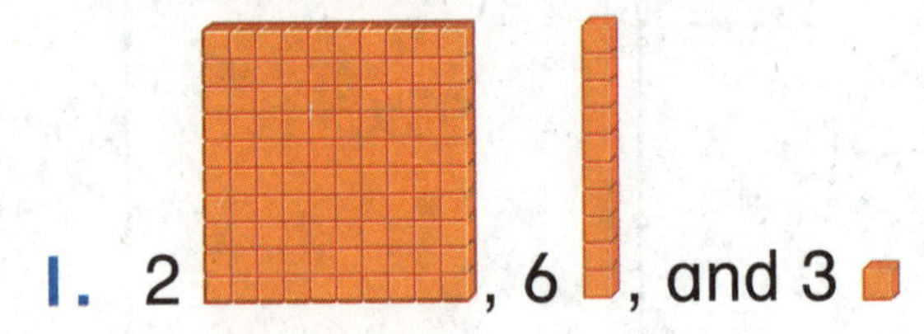

1. 2 hundreds, 6 tens, and 3 ones

hundreds	tens	ones
2	6	3

263

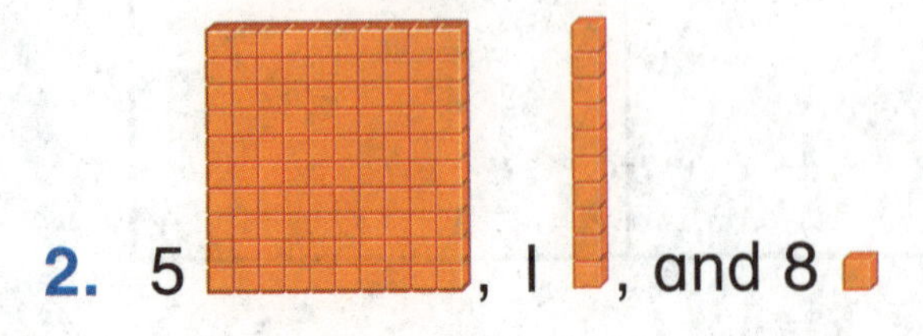

2. 5 hundreds, 1 ten, and 8 ones

hundreds	tens	ones
5	1	8

518

Talk Math What is the value of the 1 in 712, in 165, and in 381?

Name ....

CCSS

On My Own

Write how many hundreds, tens, and ones. Then write the number.

3. 3 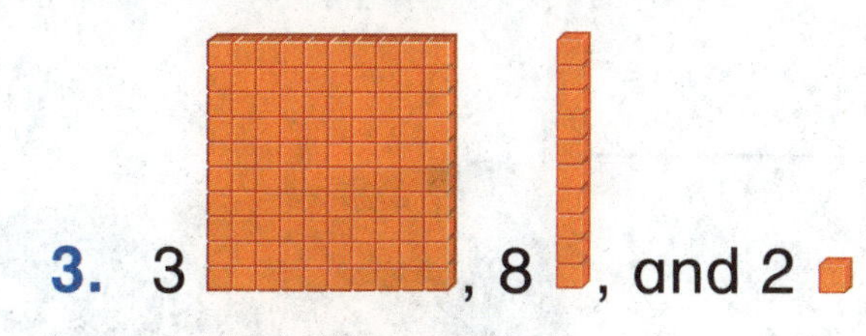 , 8, and 2

hundreds	tens	ones
3	8	2

382

4. 6 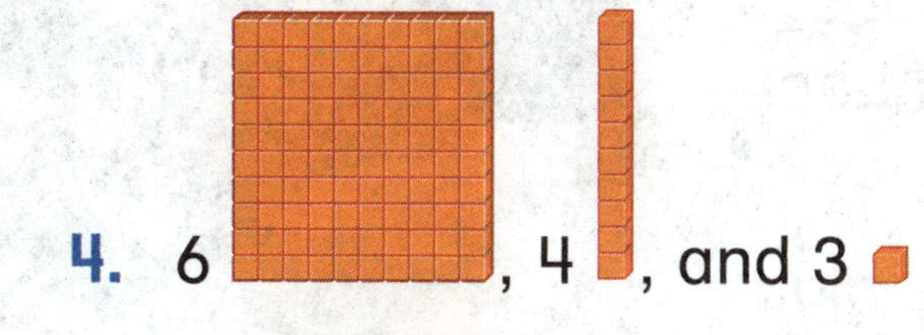 , 4, and 3

hundreds	tens	ones
6	4	3

643

5. 7 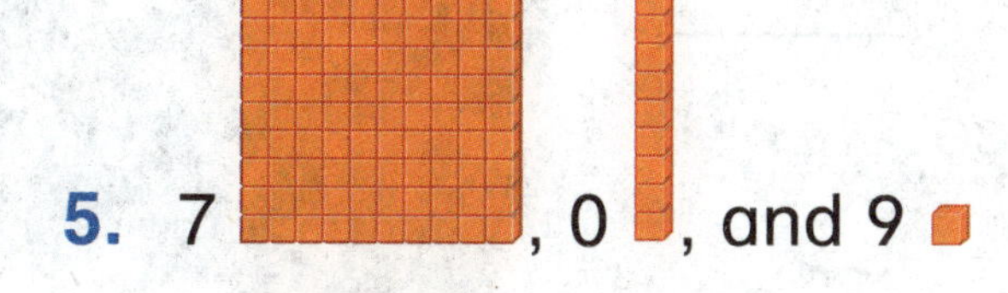 , 0, and 9

hundreds	tens	ones
7	0	9

709

6. 7 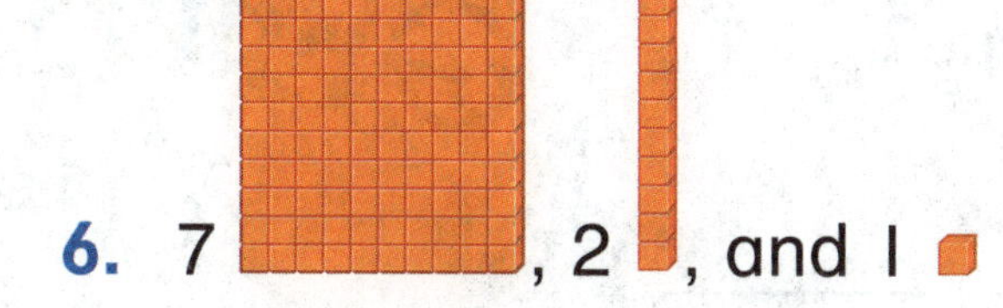 , 2, and 1

hundreds	tens	ones
7	2	1

721

Write how many hundreds, tens, and ones.

7. 684 = _______ hundreds, _______ tens, and _______ ones

8. 805 = _______ hundreds, _______ tens, and _______ ones

9. 290 = _______ hundreds, _______ tens, and _______ ones

Problem Solving

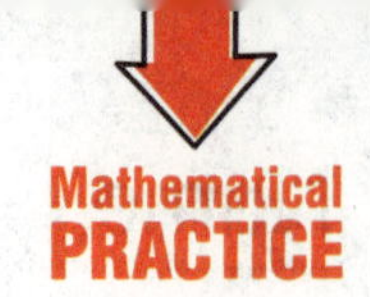

10. This number has 4 hundreds and 2 ones. The number of tens is the digit between the number of ones and the number of hundreds. What is the number?

______

You can view the largest, most complete T. Rex at The Field Museum in Chicago, Illinois. It has more than 200 bones. Look at the number 200.

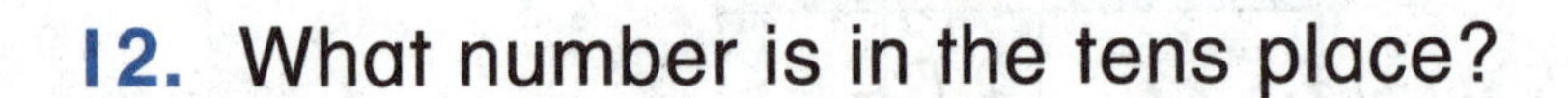

11. What number is in the hundreds place?

______

12. What number is in the tens place?

______

13. What number is in the ones place?

______

Write Math What do hundreds, tens, and ones tell you about a number?

____

____

____

Name

Number and Operations in Base Ten
2.NBT.1, 2.NBT.1a, 2.NBT.1b, 2.NBT.3

CCSS

My Homework

Lesson 2
Hundreds, Tens, and Ones

Homework Helper

Need help? connectED.mcgraw-hill.com

Use hundreds, tens, and ones to show a number.

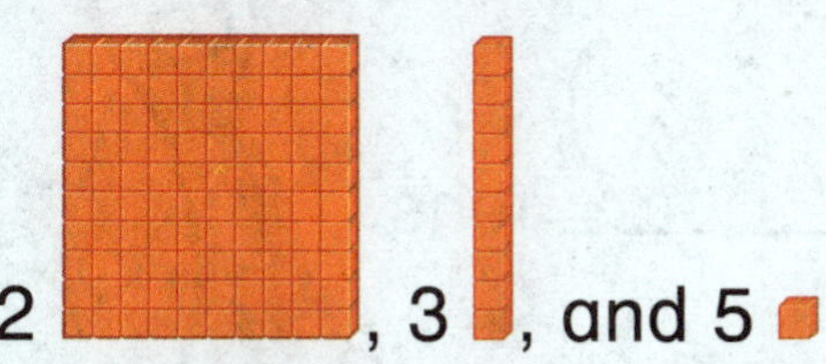

2 [hundreds], 3 [tens], and 5 [ones]

hundreds	tens	ones
2	3	5

Practice

Write how many hundreds, tens, and ones. Then write the number.

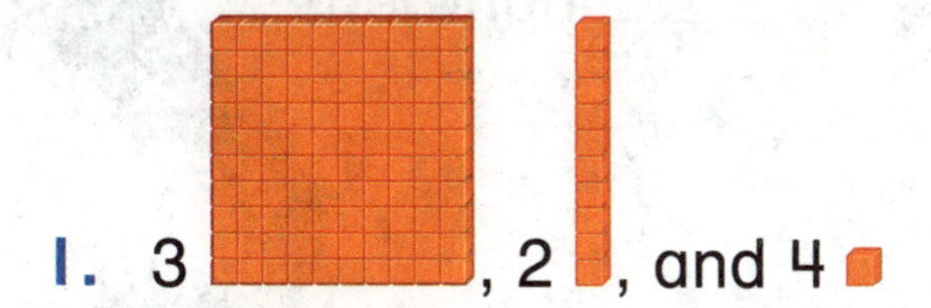

1. 3 [hundreds], 2 [tens], and 4 [ones]

hundreds	tens	ones
3	2	4

324

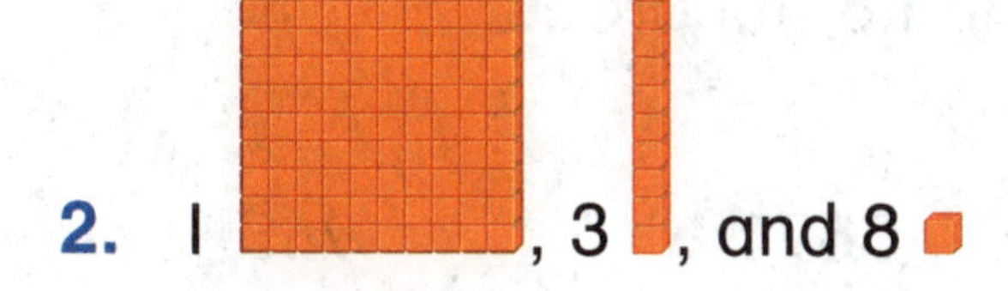

2. 1 [hundreds], 3 [tens], and 8 [ones]

hundreds	tens	ones
1	3	8

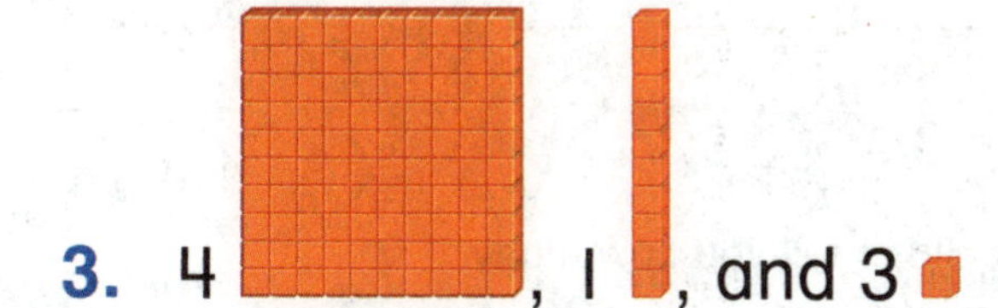

3. 4 [hundreds], 1 [tens], and 3 [ones]

hundreds	tens	ones
4	1	3

413

Write how many hundreds, tens, and ones.

4. 536 = ______ hundreds, ______ tens, and ______ ones

5. 295 = ______ hundreds, ______ tens, and ______ ones

Solve. Draw base-ten blocks to model.

Hey! I'm stuck!

6. Percy has 342 pencils.
How many tens does he have?

______ tens

7. Luis collected 613 beads.
How many hundreds does he have?

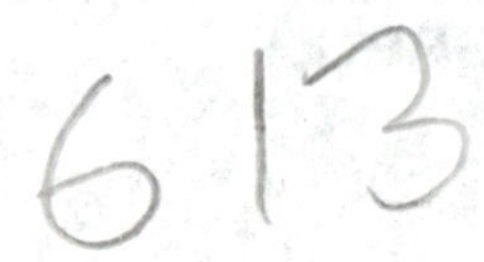

______ hundreds

Test Practice

8. A number has 6 tens, 9 ones, and 5 hundreds.
What number is it?

695	596	569	965
◉	○	○	○

Math at Home Write the number 647. Ask your child to tell you how many hundreds, tens, and ones are in the number.

Name ....

Place Value to 1,000

Lesson 3

ESSENTIAL QUESTION
How can I use place value?

Explore and Explain

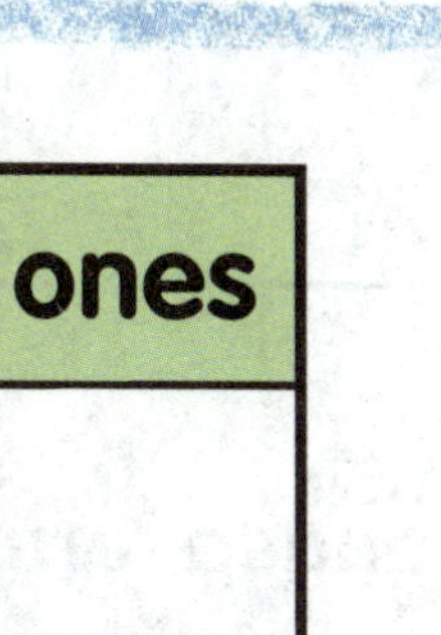

hundreds	tens	ones

_____ + _____ + _____ = _____

Teacher Directions: Roll a red number cube. Write that number in the hundreds place. Roll again. Write that number in the tens place. Roll again. Write that number in the ones place. Write the value of each number and their sum when added together.

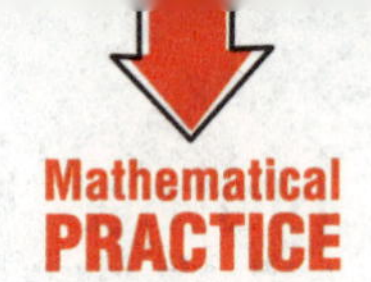

See and Show

Place value tells the value of a **digit** in a number. To write a number in **expanded form**, show the value of each digit.

hundreds	tens	ones
2	1	3

Helpful Hint
When you write a number in expanded form, you show the value of each digit.

2 hundreds 1 ten 3 ones

200 + 10 + 3 = 213

Write each number in expanded form. Then write the number.

1. 1 hundred, 4 tens, and 9 ones

_____ + _____ + _____ = _____

2. 3 hundreds, 2 tens, and 5 ones

_____ + _____ + _____ = _____

3. 5 hundreds, 7 tens, and 1 one

_____ + _____ + _____ = _____

How are 562 and 265 the same? How are they different?

Name ....

CCSS

On My Own

Write each number in expanded form. Then write the number.

4. 7 hundreds, 0 tens, and 2 ones

_______ + _______ + _______ = _______

5. 4 hundreds, 2 tens, and 1 one

_______ + _______ + _______ = _______

Circle the value of the green digit.

6. 965

900 90 9

7. 673

300 30 3

Write the number.

8. 800 + 30 + 3 = _______

9. 200 + 90 + 0 = _______

10. 700 + 10 + 9 = _______

11. 900 + 80 + 9 = _______

Write the number in expanded form.

12. 254 = _______ + _______ + _______

13. 526 = _______ + _______ + _______

Problem Solving

14. The number of raisins in a box has 4 in the ones place, 2 in the hundreds place, and 5 in the tens place. How many raisins are in the box?

_______ raisins

15. Sharon has a pile of 100 erasers, a pile of 20 erasers, and a pile of 5 erasers. How many erasers does Sharon have in all?

_______ erasers

16. Luis wrote a story. The number of pages in his story has 1 hundred, 7 tens, and 9 ones. How many pages are in his story?

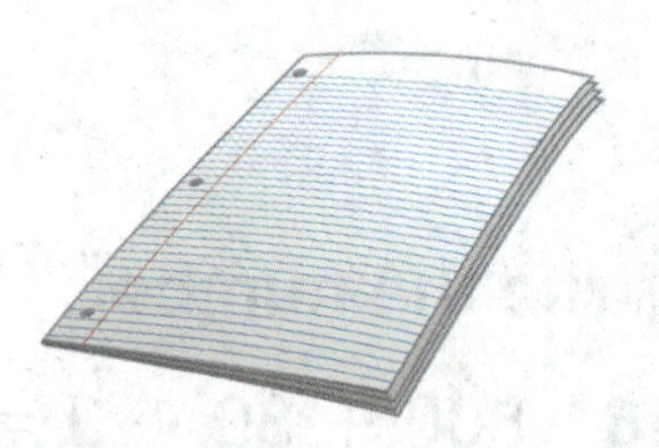

_______ pages

HOT Problem A number has three digits. Which digit has the greatest value and why?

____

____

____

Name

My Homework

Lesson 3
Place Value to 1,000

Homework Helper

eHelp Need help? connectED.mcgraw-hill.com

You can write a number in expanded form by writing the value of each digit.

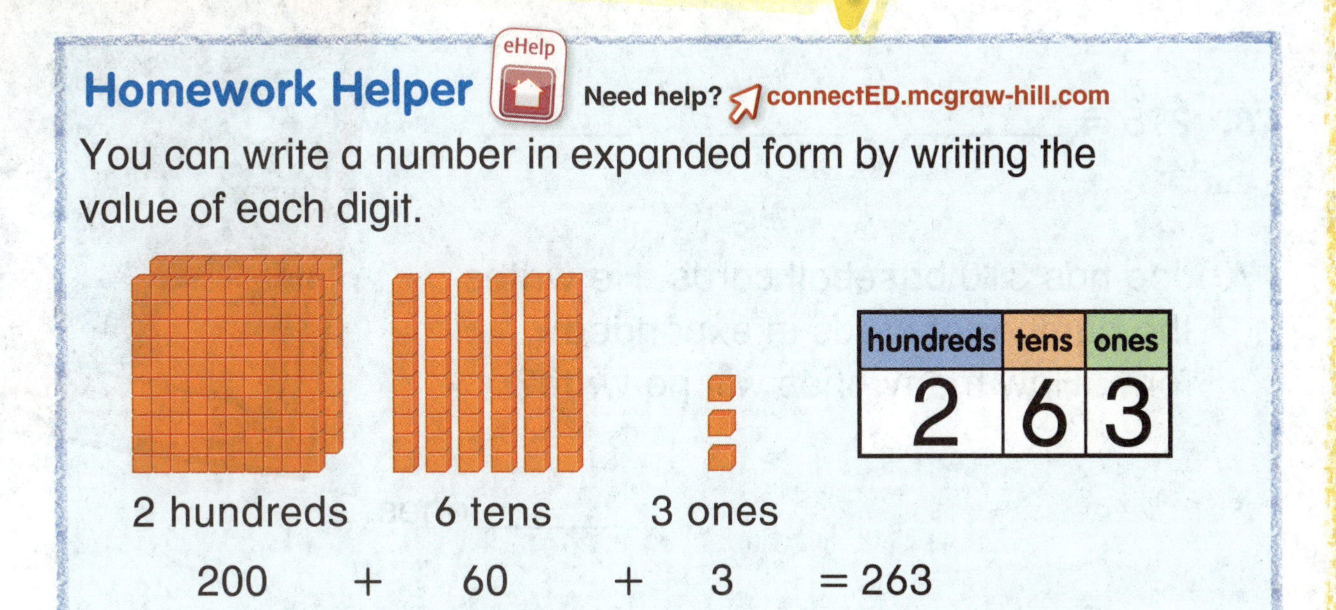

Practice

Write each number in expanded form. Then write the number.

1. 2 hundreds, 6 tens, and 8 ones

______ + ______ + ______ = ______

2. 3 hundreds, 1 ten, and 5 ones

______ + ______ + ______ = ______

3. 6 hundreds, 0 tens, and 2 ones

______ + ______ + ______ = ______

Write the number in expanded form.

4. 637 = ______ + ______ + ______

5. 742 = ______ + ______ + ______

6. 295 = ______ + ______ + ______

Here's what I'll look like on my baseball card!

7. Joe has 310 baseball cards. He writes the number of cards in expanded form. How many ones will he write?

______ ones

8. Trey reads that 482 people went to the baseball game. How can Trey show this number of people in expanded form?

______ + ______ + ______

Vocabulary Check

Circle the numbers that show the word below.

9. expanded form 300 + 42 300 + 40 + 2 3 + 4 + 2

Math at Home Have your child tell you a three-digit number. Then ask your child to tell you the value of the first digit.

Name ....

Check My Progress

Vocabulary Check

Draw lines to match.

1. hundreds

2. expanded form

3. place value

4. digit

500 + 60 + 3

hundreds	tens	ones
5	6	3

3

563
↑

Concept Check

Write how many tens and hundreds.

5. 900 ones = _______ tens = _______ hundreds

6. 600 ones = _______ tens = _______ hundreds

Write how many hundreds, tens, or ones.

7. 300 = _______ ones

8. 400 = _______ tens

Write how many hundreds, tens, and ones.
Then write the number.

9. 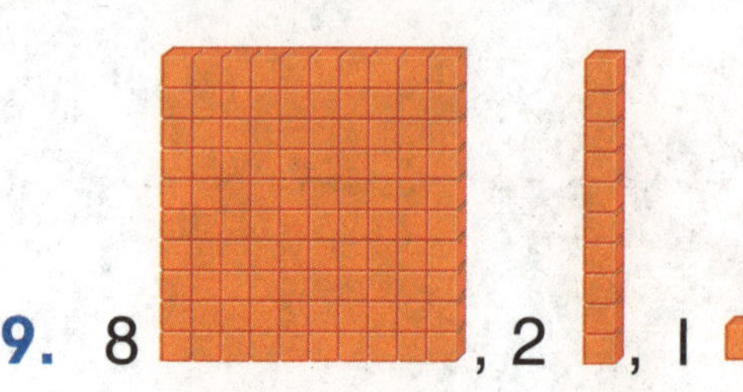

hundreds	tens	ones

______

Write each number in expanded form.
Then write the number.

10. five hundred sixty-six

______ + ______ + ______ = ______

11. nine hundred ten

______ + ______ + ______ = ______

12. Janie is thinking of a number. It has 8 ones and 9 tens. It has the same amount of hundreds as tens. What is the number?

______

Test Practice

13. A number has a 5 in the tens place. The number of ones is 3 less than the number of tens. The number of hundreds is 1 less than the number of ones. What is the number?

512 ○ 215 ○ 152 ○ 125 ○

Name ____________________

Problem Solving

STRATEGY: Use Logical Reasoning

Lesson 4

ESSENTIAL QUESTION
How can I use place value?

Watch

Eva's house number has three digits. The sum of the digits is 7. Two of the digits are even. The middle digit is 1. The first digit is less than the last digit. What is Eva's house number?

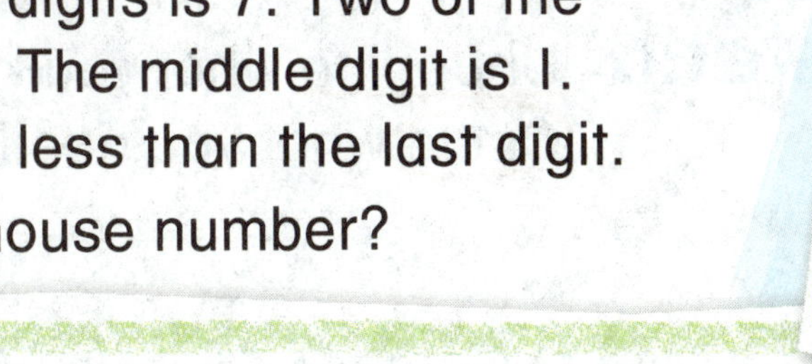

1 Understand

Underline what you know.
Circle what you need to find.

2 Plan

How will I solve the problem?

3 Solve

Use logical reasoning.

First digit plus last digit equals __6__.

__2__ and __4__ are even and equals __6__.

__2__ is less than __4__.

Eva's house number is __214__.

4 Check

Is my answer reasonable? Explain.

Practice the Strategy

The number of dogs at the shelter is a two-digit, odd number. It is more than 80. The tens digit is less than the ones digit. The sum of the digits is 17.

How many dogs are at the shelter?

1 Understand Underline what you know.
Circle what you need to find.

2 Plan How will I solve the problem?

3 Solve I will...

_______ dogs

4 Check Is my answer reasonable? Explain.

Good night!

Name

CCSS

Apply the Strategy

1. A number has three digits. The digit in the hundreds place is the difference of 4 and 2. The digit in the tens place is the sum of 2 and 3. The digit in the ones place is 2 more than the digit in the tens place. What is the number?

______

2. Sylvia writes this number: three ones, two tens, and seven hundreds. What number is it?

______

3. Ruby is trying to solve a number riddle. The digit in the tens place is less than 2. It is an odd number. The number in the hundreds place is greater than 8. The number in the ones place is the sum of 6 and 3. What is the number?

I LOVE riddles!

______

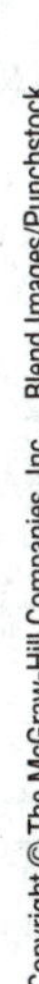

Review the Strategies

Choose a strategy

- Make a model.
- Write a number sentence.
- Find a pattern.

4. Paco puts pennies into three groups. He counts 80 pennies in the first group, 3 pennies in the second group, and 500 pennies in the third group. How many pennies does he have?

_______ pennies

5. Ella's family is going to visit her cousins. They will drive for 3 days. Each day they will drive 60 miles. How many miles will they drive in all?

_______ miles

6. Jamie ate 20 cherries on Monday. She ate 20 cherries on Tuesday. She ate 20 cherries on Wednesday. If she continues, how many cherries will Jamie have eaten by Thursday?

_______ cherries

Name

My Homework

Lesson 4

Problem Solving: Use Logical Reasoning

eHelp

Pat's Fruit Stand has 534 peaches for sale. Pat sells 30 peaches this afternoon. How many peaches are left to sell?

Peachy!

1 Understand Underline what you know.
Circle what you need to find.

2 Plan How will I solve the problem?
I will use logical reasoning.

3 Solve 534 = 500 + 30 + 4
30 peaches were sold.
504 = 500 + 30 + 4
There are 504 peaches left.

4 Check Is my answer reasonable?

Underline what you know. Circle what you need to find.

1. Lucy is thinking of a number. The number is greater than two hundred twenty-five. Her number is less than 2 hundreds, 2 tens, and 7 ones. What is Lucy's number?

____

2. The number of blocks has 9 in the ones place. The number in the hundreds place is one more than the number in the tens place. Those two numbers equal 11. How many blocks are there?

_______ blocks

3. Ralph has 957 race car stickers. He gave 10 to Ken. He gave 100 to Matt. How many stickers does Ralph have left?

_______ stickers

4. My class collected 389 milk jugs. We used 40 jugs. We gave 30 milk jugs to the first graders. How many milk jugs did we have left?

Math at Home Ask your child to write the number 961 on a piece of paper. Tell him or her to write the number that is one less than that number on the paper in expanded form.

_______ milk jugs

Name ______________________________

Read and Write Numbers to 1,000

Lesson 5

ESSENTIAL QUESTION
How can I use place value?

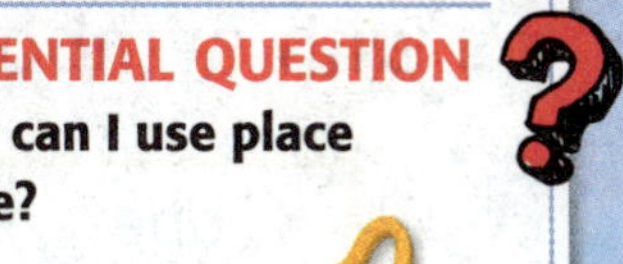

Explore and Explain

101	102	103	104	105	106	107	108	109	110
111	112	113	114		116	117	118	119	120
121	122	123	124	125	126	127	128	129	130
131		133	134	135	136	137	138	139	140
141	142	143	144	145	146	147	148	149	150
151	152	153	154	155	156		158	159	160
161	162	163	164	165	166	167	168	169	170
171	172	173	174	175	176	177		179	180
181	182	183	184	185	186	187	188	189	190
191	192	193	194	195	196	197	198		200

Teacher Directions: Fill in the missing numbers. Read the numbers from 101 to 150. Color those numbers yellow. Read the numbers from 151 to 200. Color those numbers green.

See and Show

Mathematical PRACTICE

Helpful Hint
One thousand is equal to ten hundreds. We write this as 1,000.

You can read and write numbers with symbols and words.

1 one	11 eleven	10 ten	100 one hundred
2 two	12 twelve	20 twenty	200 two hundred
3 three	13 thirteen	30 thirty	300 three hundred
4 four	14 fourteen	40 forty	400 four hundred
5 five	15 fifteen	50 fifty	500 five hundred
6 six	16 sixteen	60 sixty	600 six hundred
7 seven	17 seventeen	70 seventy	700 seven hundred
8 eight	18 eighteen	80 eighty	800 eight hundred
9 nine	19 nineteen	90 ninety	900 nine hundred
			1,000 one **thousand**

Read five hundred thirty-eight.

Write the number. 538

Read the number. Write the number.

1. thirty-eight ________

2. one hundred twenty-one ________

Write the number in words. Read the number.

3. 710 ________________________________

4. 900 ________________________________

Explain how you would write 62 and 602 using words.

Name ____________________

CCSS

Are you a Kangaroo or a squirrel?!

On My Own

Read the number.
Write the number.

5. one hundred ninety-nine

____

6. seven hundred twenty-eight

____

7. three hundred ten

____

8. two hundred eighty-five

____

9. nine hundred seventy-seven

____

10. four hundred sixty-four

____

Write the number in words. Read the number.

11. 1,000 ______________________________

12. 718 ______________________________

13. 614 ______________________________

14. 244 ______________________________

15. 321 ______________________________

Problem Solving

Mathematical PRACTICE

16. There are 119 frogs in the pond. Write the number of frogs in the pond in words.

____

17. Amber is thinking of this number word: two hundred fifty-six. What is the number?

____

18. There are four hundred twenty-seven geese in the lake. Write the number of geese in the lake.

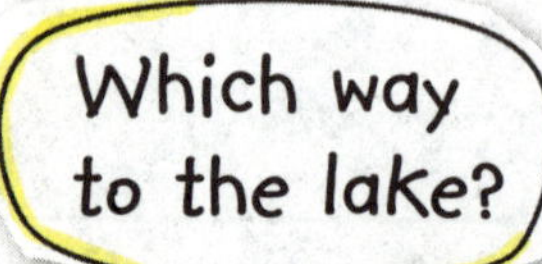

__________ geese

Write Math Why is it important to know numbers and number names?

____

____

____

Name ....

My Homework

Lesson 5
Read and Write Numbers to 1,000

Homework Helper

Need help? connectED.mcgraw-hill.com

Nine hundred fifty-four in numbers = 954

467 = four hundred sixty-seven

Practice

Read the number. Write the number.

1. two hundred thirty-five

   ______

2. six hundred seventy-two

   ______

3. one hundred forty-six

   ______

4. three hundred twenty-one

   ______

Write the number in words.
Read the number.

5. 682 ____________________________________

6. 431 ____________________________________

Solve. Draw base-ten blocks if needed.

7. There are 429 students at Linden School. Cora wants to write the number in words to put in a newsletter. What should she write?

_________________________________ students

8. Marco lives at nine hundred thirty-one Maple Street. Write the number.

____

9. Rani knows that there are three hundred sixty-five days in one year. How can she write this number?

_________ days

Vocabulary Check

10. Circle the number that shows **thousands**.

1,365   1,000   3,651

Math at Home Say three hundred forty-seven and have your child write the number.

Name ....

Count by 5s, 10s, and 100s

Lesson 6

ESSENTIAL QUESTION
How can I use place value?

Explore and Explain

201	202	203	204	205	206	207	208	209	210
211	212	213	214	215	216	217	218	219	220
221	222	223	224	225	226	227	228	229	230
231	232	233	234	235	236	237	238	239	240
241	242	243	244	245	246	247	248	249	250
251	252	253	254	255	256	257	258	259	260
261	262	263	264	265	266	267	268	269	270
271	272	273	274	275	276	277	278	279	280
281	282	283	284	285	286	287	288	289	290
291	292	293	294	295	296	297	298	299	300

Teacher Directions: Skip count by 5. Color those numbers yellow. Skip count by 10. Circle those numbers red. If the chart continued, and you counted by hundreds, what would be the next number you would say?

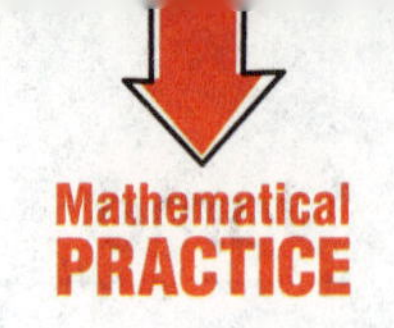

See and Show

Helpful Hint
Patterns can also be 5 less, 10 less, or 100 less.

Number patterns can help you count. In these patterns, each number is more.

Count by **fives**. Each number is ___5___ **more**.

Count by **tens**. Each number is ___10___ **more**.

ENGINE HOUSE #1

Count by **hundreds**. Each number is ___100___ **more**.

Write the missing numbers. Then write the counting pattern.

1. 340, 350, 360, ____, 380

   The pattern is __________.

2. 575, 580, ____, 590, ____

   The pattern is __________.

3. 941, 841, ____, 641, 541

   The pattern is __________.

4. 680, 675, ____, ____, 660

   The pattern is __________.

Talk Math How can you tell if a number pattern is counting by hundreds?

Name

I will make the missing numbers appear!

CCSS

On My Own

Write the missing numbers.
Then write the counting pattern.

5. 500, 510, ___, 530, ___

The pattern is _________.

6. 310, ___, 320, ___, 330

The pattern is _________.

7. 800, 790, 780, ___, 760

The pattern is _________.

8. 655, ___, 455, 355, ___

The pattern is _________.

9. ___, 486, 586, 686, ___

The pattern is _________.

10. 500, 600, 700, ___, 900

The pattern is _________.

11. 234, ___, ___, 264, 274

The pattern is _________.

12. ___, 515, 510, 505, ___

The pattern is _________.

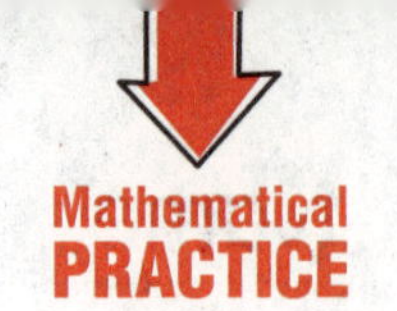

Problem Solving

13. Paul counts by hundreds. He starts with the number 123. Write the numbers Paul counts.

123, ______, ______, ______, ______, ______

14. Ali counts by tens. She starts with 325. Ali counts 325, 335, 345, 365, 385. What numbers did Alli leave out?

What's next?

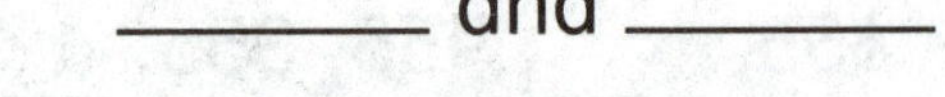

15. Write the numbers you would say if you count backwards by fives. Start with 230 and end at 200.

230, ______, ______, ______, ______, ______, 200

Felicia counts from 300 to 400. Would it be faster to count by 5s or by 10s? Explain.

____

____

____

Name ....

My Homework

Lesson 6
Count by 5s, 10s, and 100s

Homework Helper

Need help? connectED.mcgraw-hill.com

Number patterns can help you count.

Count by fives: 40, 45, 50, 55, 60, 65

Count by tens: 120, 130, 140, 150, 160

Count by hundreds: 200, 300, 400, 500, 600

Practice

Write the missing numbers. Then write the counting pattern.

1. 1,000, 995, ____, 985, ____

   The pattern is ________.

2. 524, ____, 544, 554, ____

   The pattern is ________.

3. ____, 283, 383, ____, 583

   The pattern is ________.

4. ____, 843, 743, 643, ____

   The pattern is ________.

5. 420, 430, ____, 450, ____

   The pattern is ________.

6. 525, ____, 535, ____, 545

   The pattern is ________.

Continue the counting pattern.

7. 835, 840, 845, ______, ______, ______

8. 410, 420, 430, ______, ______, ______

9. 900, 800, 700, ______, ______, ______

---

10. Shari counts: 169, 159, 149, 139, 129. Shari wants Miguel to guess her counting pattern. What should Miguel guess?

The pattern is ____________.

Test Practice

11. Mark the number pattern that shows 10 less.

820, 830, 840 ○	980, 970, 960 ○
923, 922, 921 ○	400, 500, 600 ○

Math at Home Pick a three-digit number. Ask your child to count by ones, tens, and/or hundreds.

Name ....

Compare Numbers to 1,000

Lesson 7

ESSENTIAL QUESTION
How can I use place value?

Explore and Explain

How much are you worth?

Teacher Directions: Use base-ten blocks to show 331 and 208 on the mat. Place them on the appropriate side of the symbol. Draw the base-ten blocks.

See and Show

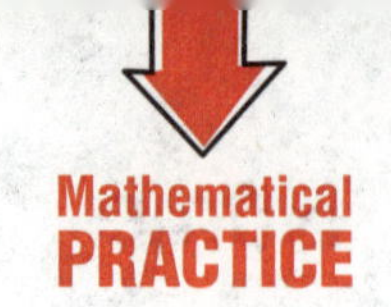

Use these steps to compare numbers.

Step 1 Line up the numbers by place value in the chart.

Step 2 **Compare** the digits in the greatest place value first. If they are the same, compare digits in the next place value.

greater than

hundreds	tens	ones
6	1	8
6	3	4

634 > 618

6**3**4 is *greater than* 6**1**8.

less than

hundreds	tens	ones
8	4	6
8	2	3

823 < 846

8**2**3 is *less than* 8**4**6.

equal to

hundreds	tens	ones
9	7	8
9	7	8

978 = 978

978 is *equal to* **978.**

Compare. Write >, <, or =.

1. 142 ◯ 124
2. 253 ◯ 257
3. 313 ◯ 313
4. 842 ◯ 795
5. 694 ◯ 694
6. 203 ◯ 153
7. 100 ◯ 1,000
8. 999 ◯ 99
9. 133 ◯ 133
10. 743 ◯ 734
11. 861 ◯ 871
12. 542 ◯ 452

Talk Math Explain how to compare 567 and 576.

Name ______________________________

On My Own

Helpful Hint
Compare the digits in the greatest place value first.

Compare. Write >, <, or =.

13. 150 ◯ 150

14. 132 ◯ 213

15. 689 ◯ 627

16. 425 ◯ 425

17. 907 ◯ 899

18. 533 ◯ 533

19. 207 ◯ 210

20. 697 ◯ 667

21. 411 ◯ 421

22. 619 ◯ 621

23. 729 ◯ 729

24. 325 ◯ 300

25. 332 ◯ 335

26. 984 ◯ 894

Circle the number that is less than the purple number.

27. 568

588 464

28. 409

410 406

Circle the number that is greater than the purple number.

29. 311

313 211

30. 653

651 655

Problem Solving

31. I am greater than 3 hundreds, 2 tens, and 2 ones. I am less than 3 hundreds, 2 tens, 4 ones. What number am I?

____

32. Ravi is making cookies. He wants to have two equal groups of cookies. He made 45 cookies for one group. How many cookies does Ravi need for the other group?

_______ cookies

33. Demi collected 124 seashells. Eva collected 142 seashells. Who collected less seashells?

____

Write Math Explain how you decide which number is greater than the other.

____

____

____

Name ....

My Homework

Lesson 7

Compare Numbers to 1,000

Homework Helper

Need help? connectED.mcgraw-hill.com

greater than

hundreds	tens	ones
5	4	6
4	6	2

546 > 462

less than

hundreds	tens	ones
3	2	5
5	1	8

325 < 518

equal to

hundreds	tens	ones
2	5	8
2	5	8

258 = 258

Practice

Compare. Write >, <, or =.

1. 415 ◯ 451
2. 623 ◯ 678
3. 730 ◯ 830
4. 375 ◯ 375
5. 549 ◯ 560
6. 258 ◯ 239
7. 109 ◯ 111
8. 382 ◯ 379
9. 445 ◯ 545
10. 272 ◯ 275
11. 818 ◯ 816
12. 357 ◯ 357

Solve to answer each question below.

13. Uri has 529 bugs in his collection.
Elena has 513 bugs in her collection.
Who has a greater number of bugs?

____

14. Seth saves 347 bottle caps.
Jorge saves 345. Who saves
the greater number of bottle caps?

____

15. Elisa has 125 paper dolls in her collection.
Katherine has 138 paper dolls in her
collection. Who has less dolls?

____

Vocabulary Check

Draw lines to match.

16. **greater than**	=
17. **less than**	>
18. **equal to**	<

Math at Home Ask your child to name numbers that are greater than, less than, and equal to 807.

Name ....

My Review

Chapter 5
Place Value to 1,000

Vocabulary Check 

Draw lines to match.

1. **expanded form**	498
2. **hundreds**	>
3. **equal to**	1,000
4. **greater than**	=
5. **less than**	700 + 30 + 6
6. **thousand**	<

Concept Check 

Write how many hundreds, tens, and ones.

7. 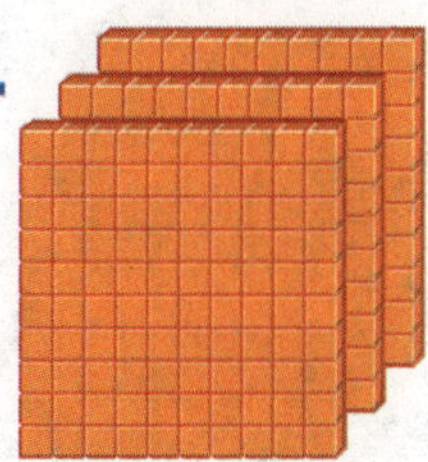

_____ hundreds = _____ tens = _____ ones

Write how many hundreds, tens, and ones.
Then write the number.

8. 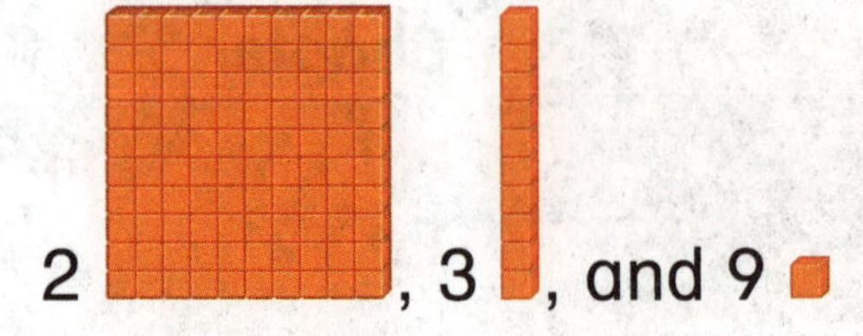
2 [hundreds], 3 [tens], and 9 [ones]

hundreds	tens	ones

______

Write the number in expanded form.
Then write the number.

9. 5 hundreds, 6 tens, and 0 ones

______ + ______ + ______ = ______

10. 8 hundreds, 7 tens, and 8 ones

______ + ______ + ______ = ______

Write the number.

11. six hundred twenty-nine

______

12. five hundred one

______

Compare. Write >, <, or =.

13. 100 ◯ 1,000

14. 468 ◯ 406

15. 711 ◯ 711

16. 394 ◯ 294

Name ....

CCSS

Problem Solving

17. Malika has 9 sheets of stickers. Tonya has 7 sheets of stickers. 100 stickers are on each sheet. Who has more stickers?

____

18. Show another way to write six hundred twenty-three.

________ + ________ + ________

19. Sam has 998 markers. Write the number words to show how many markers.

________________________ markers

Test Practice

20. Leah has 368 sunflower seeds. How many hundreds does she have?

○ 3 ○ 6 ○ 8 ○ 368

Reflect

Chapter 5

Answering the Essential Question

Show the ways you can use place value.

ESSENTIAL QUESTION

How can I use place value?

Write the number.

seven hundred sixty-three

______

Write the number in expanded form.

233

______ + ______ + ______

Compare.

763 ◯ 637

Write the missing numbers.

560, ______, 580, ______

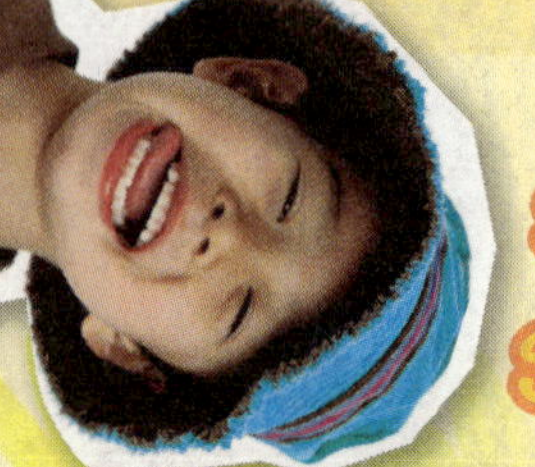

Chapter 6 Add Three-Digit Numbers

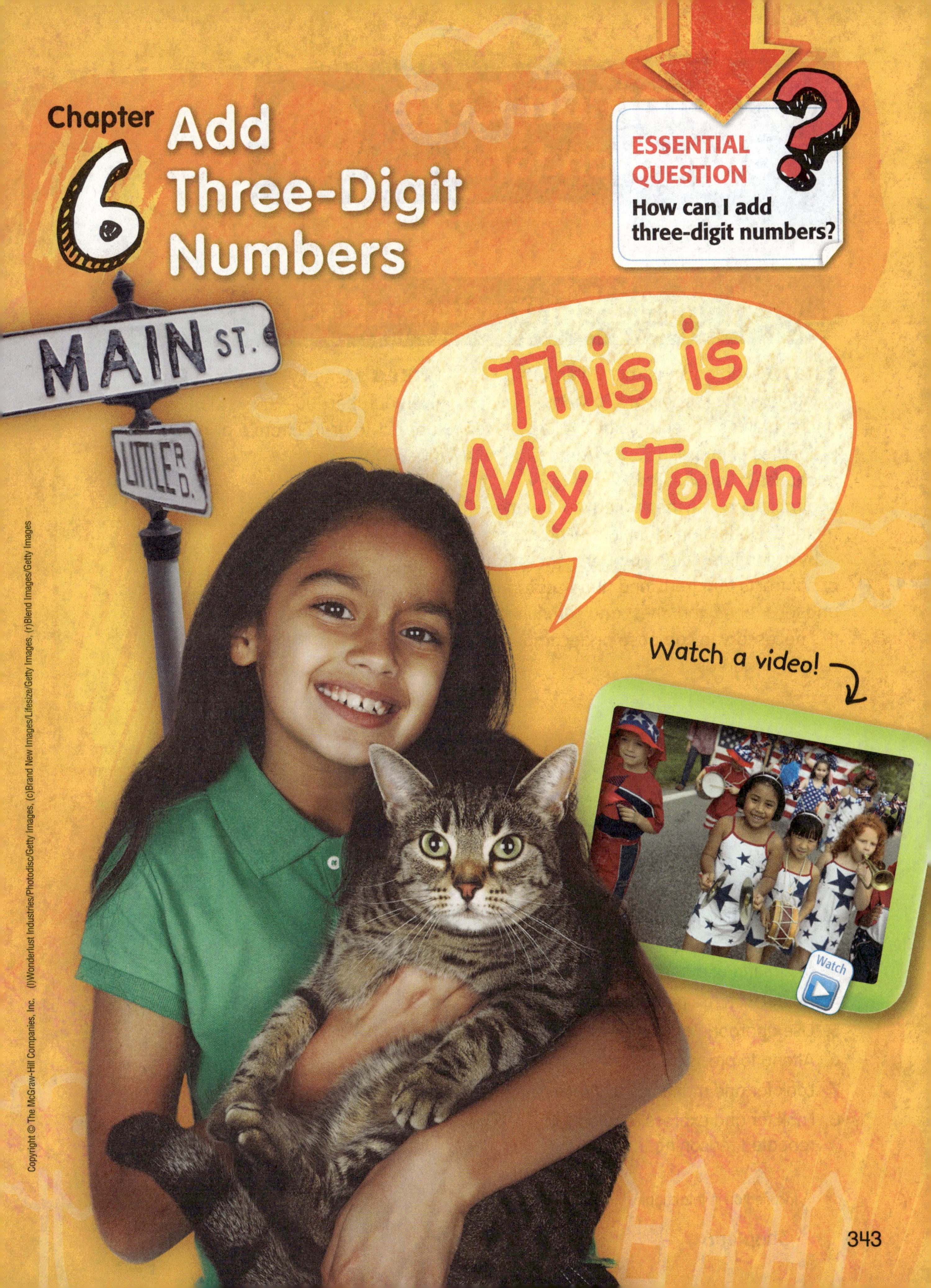

My Common Core State Standards

Number and Operations in Base Ten

2.NBT.7 Add and subtract within 1000, using concrete models or drawings and strategies based on place value, properties of operations, and/or the relationship between addition and subtraction; relate the strategy to a written method. Understand that in adding or subtracting three digit numbers, one adds or subtracts hundreds and hundreds, tens and tens, ones and ones; and sometimes it is necessary to compose or decompose tens or hundreds.

2.NBT.8 Mentally add 10 or 100 to a given number 100–900, and mentally subtract 10 or 100 from a given number 100–900.

2.NBT.9 Explain why addition and subtraction strategies work, using place value and the properties of operations.

Standards for Mathematical PRACTICE

1. Make sense of problems and persevere in solving them.
2. Reason abstractly and quantitatively.
3. Construct viable arguments and critique the reasoning of others.
4. Model with mathematics.
5. Use appropriate tools strategically.
6. Attend to precision.
7. Look for and make use of structure.
8. Look for and express regularity in repeated reasoning.

= focused on in this chapter

Name

Am I Ready?

Check ← Go online to take the Readiness Quiz

Add.

1. $\begin{array}{r} 9 \\ +\ 8 \\ \hline \end{array}$

2. $\begin{array}{r} 8 \\ +\ 5 \\ \hline \end{array}$

3. $\begin{array}{r} 40 \\ +30 \\ \hline \end{array}$

4. $\begin{array}{r} 56 \\ +28 \\ \hline \end{array}$

5. $\begin{array}{r} 8 \\ +\ 8 \\ \hline \end{array}$

6. $\begin{array}{r} 14 \\ +\ 5 \\ \hline \end{array}$

7. $\begin{array}{r} 60 \\ +30 \\ \hline \end{array}$

8. $\begin{array}{r} 83 \\ +12 \\ \hline \end{array}$

9. $\begin{array}{r} 27 \\ +18 \\ \hline \end{array}$

10. Destiny eats 21 peanuts on Friday.
She eats 29 peanuts on Saturday.
How many peanuts did she eat in all?

_______ peanuts

Shade the boxes to show the problems you answered correctly.

1	2	3	4	5	6	7	8	9	10

Name

My Math Words

Review Vocabulary

ones regroup sum tens

Use the review words to tell what the example shows about the word problem.

$$\begin{array}{r} 2\ \boxed{9} \\ +1\ \boxed{6} \\ \hline \end{array}$$

______

$$\begin{array}{r} \boxed{2}\ 9 \\ +\ \boxed{1}\ 6 \\ \hline \end{array}$$

______

There are 29 houses on my street. There are 16 houses on my friend's street. How many houses in all?

$$\begin{array}{r} \boxed{1}\ \ \\ 29 \\ +16 \\ \hline \end{array}$$

______

$$\begin{array}{r} 29 \\ +16 \\ \hline \boxed{45} \end{array}$$

______

My Vocabulary Cards

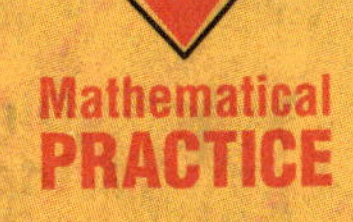

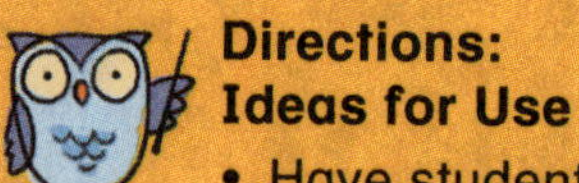

Directions:
Ideas for Use

- Have students use the blank cards to write words from previous chapters that they would like to review.
- Ask students to use the blank cards to write their own vocabulary cards.

FOLDABLES Follow the steps on the back to make your Foldable.

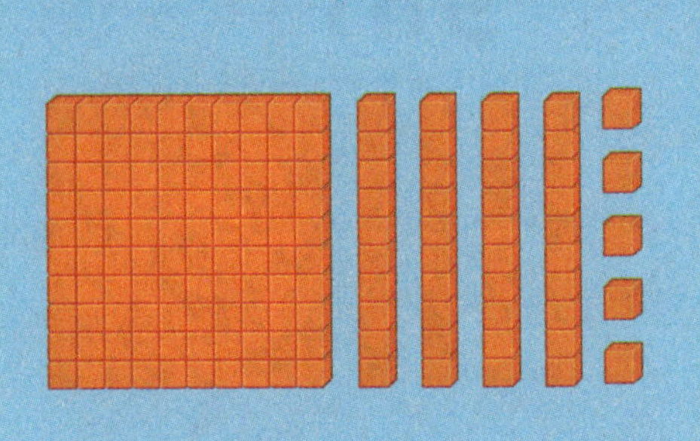

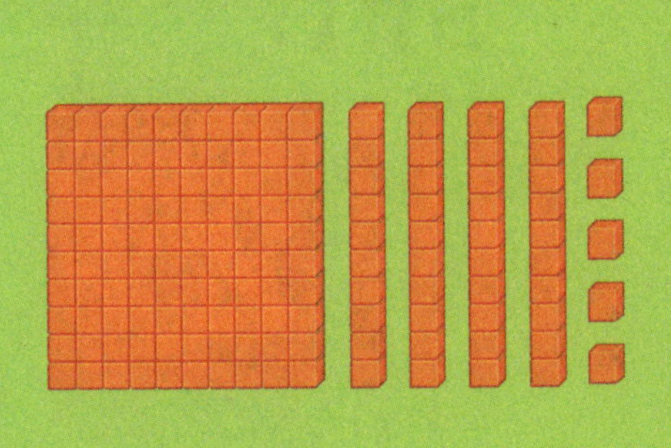

$$\begin{array}{r} 145 \\ +\ 112 \\ \hline \end{array}$$

$$\begin{array}{r} 145 \\ +\ 109 \\ \hline \end{array}$$

$$\begin{array}{r} 145 \\ +\ 155 \\ \hline \end{array}$$

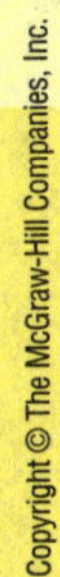

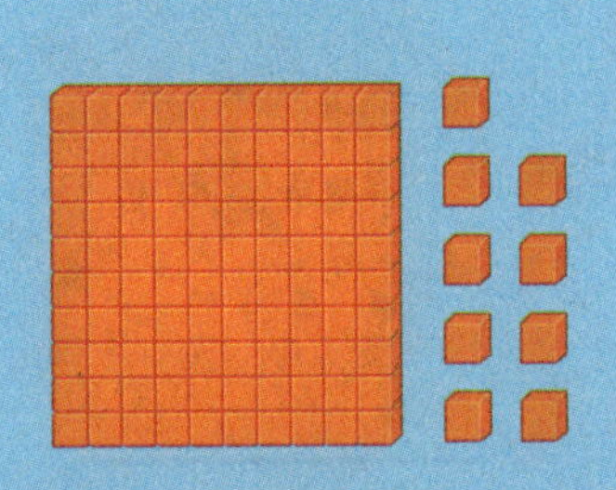

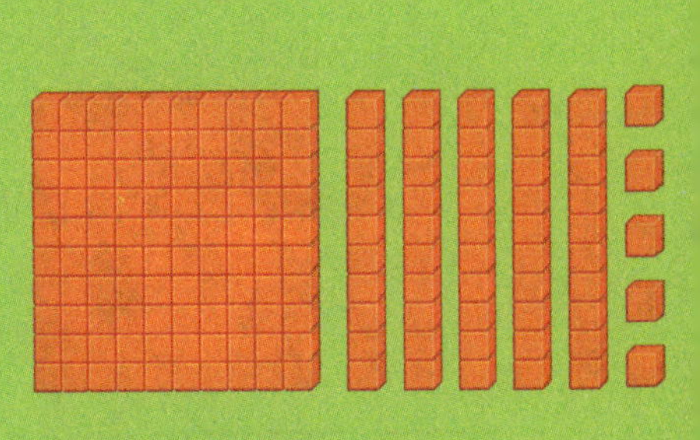

1

2
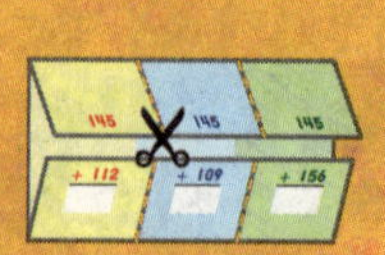

3

145

145

145

+ 155

+ 109

+ 112

Name

Make a Hundred to Add

Lesson 1

ESSENTIAL QUESTION
How can I add three-digit numbers?

Explore and Explain

Gee, how do I get to G Street?

97 + 16 = ______

______ + ______ = ______

Teacher Directions: Model using base-ten blocks. My town has 97 streets named by numbers and 16 streets named by letters of the alphabet. How many streets are named by numbers and letters of the alphabet? Take the ones from 16 and add them to 97 to make 100. Write the new problem. Add.

See and Show

You can make a hundred to help you add.

Find 199 + 43.

Take apart 43 to make 1 + 42.
Then add 1 + 199. Add 42 to that sum.

199 + 43

199 + 1 + 42

200 + 42 = 242

So, 199 + 43 = 242.

Make a hundred to add.

1. 77 + 298

_____ + _____ + 298

75 + _____ = _____

So, 77 + 298 = _____.

2. 237 + 99

_____ + _____ + 99

236 + _____ = _____

So, 237 + 99 = _____.

Talk Math Explain how you know which addend to take apart to make a hundred to add.

Name ....

Helpful Hint
Take apart a number to make a hundred. Then add.

CCSS

On My Own

Make a hundred to add.

3. 98 + 393

98 + ______ + ______

______ + 391 = ______

So, 98 + 393 = ______.

4. 45 + 199

______ + ______ + 199

44 + ______ = ______

So, 45 + 199 = ______.

5. 99 + 468

99 + ______ + ______

______ + 467 = ______

So, 99 + 468 = ______.

6. 62 + 199

______ + ______ + 199

61 + ______ = ______

So, 62 + 199 = ______.

7. 99 + 798

99 + ______ + ______

______ + 797 = ______

So, 99 + 798 = ______.

8. 94 + 198

______ + ______ + 198

92 + ______ = ______

So, 94 + 198 = ______.

Problem Solving

Mathematical PRACTICE

I'm the prettiest pig at the fair!

9. There are 299 animals at the fair. 76 more animals are brought to the fair. How many animals are at the fair now?

_______ animals

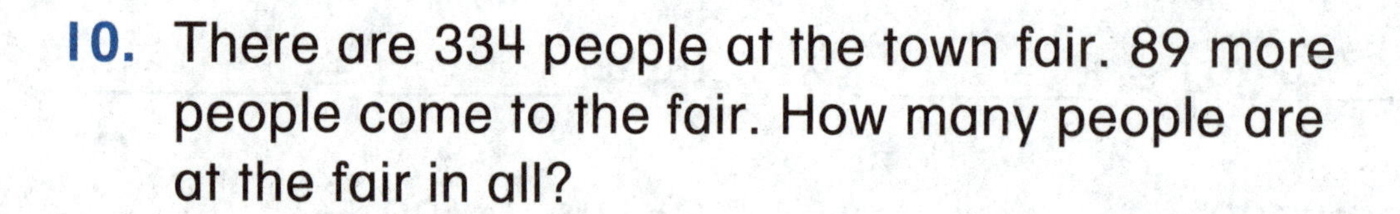

10. There are 334 people at the town fair. 89 more people come to the fair. How many people are at the fair in all?

_______ people

11. Leon is adding a number to 35. He takes apart 35 into 33 and 2 so he can use the 2 to make a hundred. Is Leon adding 199 or 198 to 35?

____

Write Math How can taking apart a number help you add to a three-digit number?

____

____

____

Name ....

My Homework

Lesson 1
Make a Hundred to Add

Homework Helper

Need help? connectED.mcgraw-hill.com

You can make a hundred to add.

Find 197 + 29.

Take apart 29 to make 26 + 3.
Then add 197 and 3 to make 200.

197 + 29

197 + 3 + 26

200 + 26 = 226

So, 197 + 29 = 226.

Helpful Hint
Take apart a number to make a hundred. Then add.

Practice

Make a hundred to add.

1. 23 + 398

_______ + _______ + 398

21 + _______ = _______

So, 23 + 398 = _______.

2. 178 + 98

_______ + _______ + 98

176 + _______ = _______

So, 178 + 98 = _______.

Make a hundred to add.

3. 77 + 196

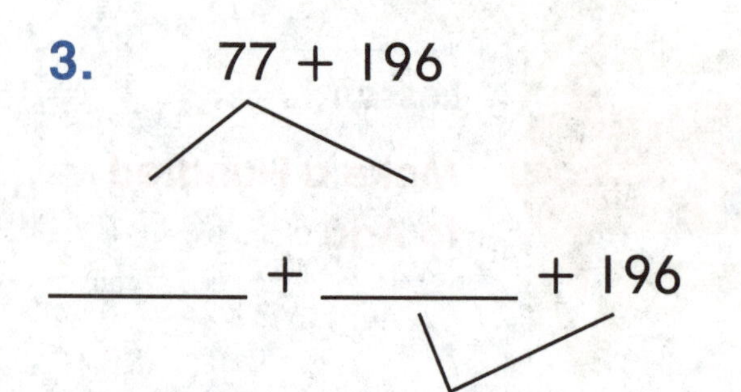

_____ + _____ + 196

73 + _____ = _____

So, 77 + 196 = _____.

4. 245 + 99

_____ + _____ + 99

244 + _____ = _____

So, 245 + 99 = _____.

5. 197 people ride the bus in the morning. 74 people ride the bus in the evening. How many people ride the bus in all?

_____ people

Test Practice

6. How could you take apart 128 to solve this addition number sentence?

128 + 94 = _____

108 + 20	100 + 28	122 + 6	120 + 8
○	○	○	○

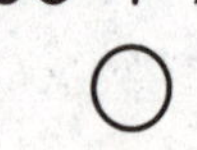

Math at Home Ask your child to solve 99 + 598 by taking apart an addend.

Name ______________________________

Add Hundreds

Lesson 2

ESSENTIAL QUESTION
How can I add three-digit numbers?

Explore and Explain

_______ + _______ = _______ patients

Teacher Directions: Model using base-ten blocks. The doctor in my town saw 200 patients last month. This month he has seen 100 patients. How many patients has the doctor seen in all? Write the addition sentence.

See and Show

You can use addition facts to add hundreds.

Find 300 + 400.

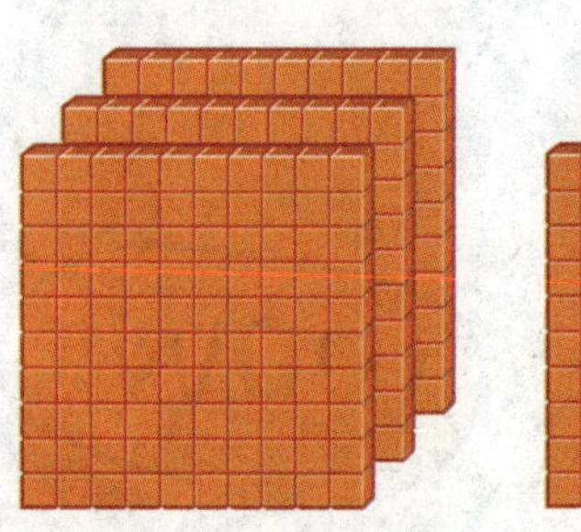

Helpful Hint
You know that
3 + 4 = 7, so
300 + 400 = 700.

  3 hundreds
+ 4 hundreds
  7 hundreds

300 + 400 = 700

Add.

1. 600 + 100 = ______
2. 300 + 300 = ______
3. 200 + 300 = ______
4. 700 + 200 = ______
5. 400 + 200 = ______
6. 800 + 100 = ______

7. 200 + 100 = ______
8. 100 + 300 = ______
9. 500 + 200 = ______

Talk Math What addition fact can help you find 600 + 100? Explain.

Name

On My Own

Add.

10. 200 + 300 = ______

11. 400 + 400 = ______

12. 700 + 100 = ______

13. 600 + 300 = ______

14. 400 + 200 = ______

15. 100 + 100 = ______

16. $\begin{array}{r} 300 \\ +\ 400 \\ \hline \end{array}$

17. $\begin{array}{r} 100 \\ +\ 800 \\ \hline \end{array}$

18. $\begin{array}{r} 200 \\ +\ 600 \\ \hline \end{array}$

19. $\begin{array}{r} 500 \\ +\ 300 \\ \hline \end{array}$

20. $\begin{array}{r} 200 \\ +\ 700 \\ \hline \end{array}$

21. $\begin{array}{r} 400 \\ +\ 300 \\ \hline \end{array}$

22. $\begin{array}{r} 600 \\ +\ 200 \\ \hline \end{array}$

23. $\begin{array}{r} 300 \\ +\ 100 \\ \hline \end{array}$

24. $\begin{array}{r} 900 \\ +\ 0 \\ \hline \end{array}$

25. $\begin{array}{r} 700 \\ +\ 100 \\ \hline \end{array}$

Here's the news...
You can add 100s!

Problem Solving

26. Ava counts 100 cars in a parking lot. She counts 200 cars in another parking lot. How many cars did Ava count in all?

_______ cars

27. A baker bakes 400 bagels on Saturday. He bakes 200 bagels on Sunday. How many bagels does he bake on those two days?

_______ bagels

28. Jan delivers 300 newspapers on Friday. She delivers 400 newspapers on Sunday. How many newspapers did she deliver in all?

_______ newspapers

HOT Problem Pam has collected 100 pull-tabs for charity. Jake has 100 of them and Pat has 200. How many do they have altogether? Tell how knowing addition facts can help you solve this problem. Solve.

____

____

____

Name

My Homework

Lesson 2
Add Hundreds

Homework Helper

Need help? connectED.mcgraw-hill.com

You can use addition facts to add hundreds.

Find 200 + 500.

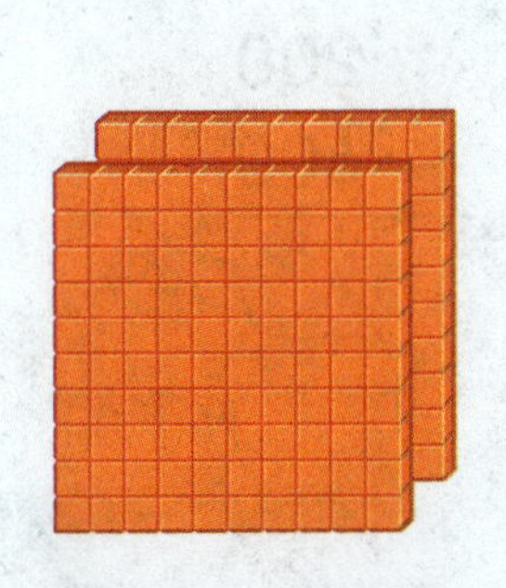

Helpful Hint
Think of addition facts you know. They can help you add hundreds.

  2 hundreds
+ 5 hundreds
  7 hundreds

200 + 500 = 700

Practice

Add.

1. 200 + 300 = ______

2. 500 + 300 = ______

3. 100 + 100 = ______

4. 600 + 100 = ______

5. $\begin{array}{r} 700 \\ +\ 200 \\ \hline \end{array}$

6. $\begin{array}{r} 800 \\ +\ 100 \\ \hline \end{array}$

7. $\begin{array}{r} 400 \\ +\ 400 \\ \hline \end{array}$

Add.

8. $100 + 200$	**9.** $500 + 100$	**10.** $200 + 400$
11. $100 + 700$	**12.** $100 + 400$	**13.** $300 + 300$
14. $700 + 200$	**15.** $500 + 400$	**16.** $200 + 600$

17. Zoe counts 300 seeds in her pumpkin.
Michael counts 500 seeds in his pumpkin.
How many seeds do they count in all?

_______ seeds

Test Practice

18. Find the sum.

$600 + 300 =$ _______

900	300	9	3
○	○	○	○

Math at Home Ask your child how knowing 4 + 5 = 9 helps him or her find 400 + 500.

Name

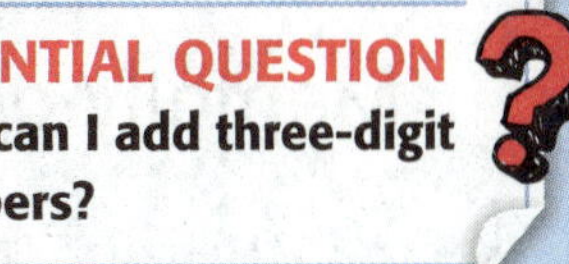

Mentally Add 10 or 100

Lesson 3

ESSENTIAL QUESTION
How can I add three-digit numbers?

Explore and Explain

Teacher Directions: Model using base-ten blocks. 126 students are waiting in line for the yellow buses and 126 students are also waiting for the green buses. 100 more students get in line for the yellow buses. 10 more students get in line for the green buses. How many students are in line for each bus now? Write the numbers.

See and Show

To add 10 or 100, think of addition facts you know.

Mentally add 100.

$$\begin{array}{r} 215 \\ +\ 100 \\ \hline \end{array}$$

Helpful Hint
You know that
2 + 1 = 3.

$$\begin{array}{r} 215 \\ +\ 100 \\ \hline 315 \end{array}$$

Mentally add 10.

$$\begin{array}{r} 250 \\ +\ 10 \\ \hline \end{array}$$

Helpful Hint
You know that
5 + 1 = 6.

$$\begin{array}{r} 250 \\ +\ 10 \\ \hline 260 \end{array}$$

Add.

1. $\begin{array}{r} 450 \\ +\ 100 \\ \hline \end{array}$

2. $\begin{array}{r} 630 \\ +\ 10 \\ \hline \end{array}$

3. $\begin{array}{r} 801 \\ +\ 100 \\ \hline \end{array}$

4. $\begin{array}{r} 115 \\ +\ 10 \\ \hline \end{array}$

5. $\begin{array}{r} 500 \\ +\ 10 \\ \hline \end{array}$

6. $\begin{array}{r} 237 \\ +\ 100 \\ \hline \end{array}$

7. 327 + 100 = ______

8. 755 + 10 = ______

Talk Math Why is it easy to mentally add 10 or 100?

Name ....

On My Own

Add.

9. $\begin{array}{r} 122 \\ +\ 100 \\ \hline \end{array}$

10. $\begin{array}{r} 378 \\ +\ 10 \\ \hline \end{array}$

11. $\begin{array}{r} 500 \\ +\ 10 \\ \hline \end{array}$

12. $\begin{array}{r} 200 \\ +\ 100 \\ \hline \end{array}$

13. $\begin{array}{r} 747 \\ +\ 100 \\ \hline \end{array}$

14. $\begin{array}{r} 800 \\ +\ 10 \\ \hline \end{array}$

15. $\begin{array}{r} 400 \\ +\ 10 \\ \hline \end{array}$

16. $\begin{array}{r} 668 \\ +\ 100 \\ \hline \end{array}$

17. $\begin{array}{r} 817 \\ +\ 100 \\ \hline \end{array}$

Find the missing addend.

18. 392 + _______ = 492

19. _______ + 10 = 825

20. 101 + _______ = 111

21. 756 + _______ = 856

22. _______ + 100 = 823

23. _______ + 10 = 846

24. 385 + _______ = 485

Problem Solving

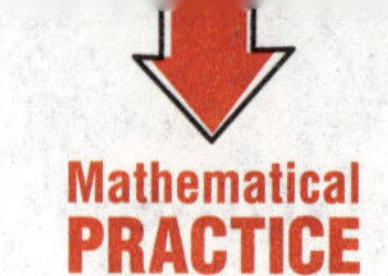

25. 448 birds are in the park. 100 more birds fly into the park. How many birds are in the park now?

_______ birds

26. A mail carrier delivered 352 packages last month. She delivered 10 more packages last week. How many packages did she deliver in all?

_______ packages

27. There are 112 sheep in a barn. More sheep come into the barn. Now there are 122 sheep. How many sheep came into the barn?

_______ sheep

Write Math Explain how to add 100 to 899.

____

____

____

Name ....

My Homework

Lesson 3
Mentally Add 10 or 100

Homework Helper

Need help? connectED.mcgraw-hill.com

Mentally add 100.

$$\begin{array}{r} 152 \\ +\ 100 \\ \hline 252 \end{array}$$

Helpful Hint
You know that
1 + 1 = 2.

Mentally add 10.

$$\begin{array}{r} 152 \\ +\ 10 \\ \hline 162 \end{array}$$

Helpful Hint
You know that
5 + 1 = 6.

Practice

Add.

1. $\begin{array}{r} 235 \\ +\ 100 \\ \hline \end{array}$

2. $\begin{array}{r} 370 \\ +\ 10 \\ \hline \end{array}$

3. $\begin{array}{r} 532 \\ +\ 100 \\ \hline \end{array}$

4. $\begin{array}{r} 365 \\ +\ 10 \\ \hline \end{array}$

5. $\begin{array}{r} 178 \\ +\ 100 \\ \hline \end{array}$

6. $\begin{array}{r} 600 \\ +\ 10 \\ \hline \end{array}$

Find the missing addend.

7. 389 + _______ = 489

8. _______ + 10 = 235

9. 634 + _____ = 644

10. 295 + _______ = 395

11. Jackson has 389 ants in his ant farm. He puts 10 more ants in the farm. How many ants are in his ant farm now?

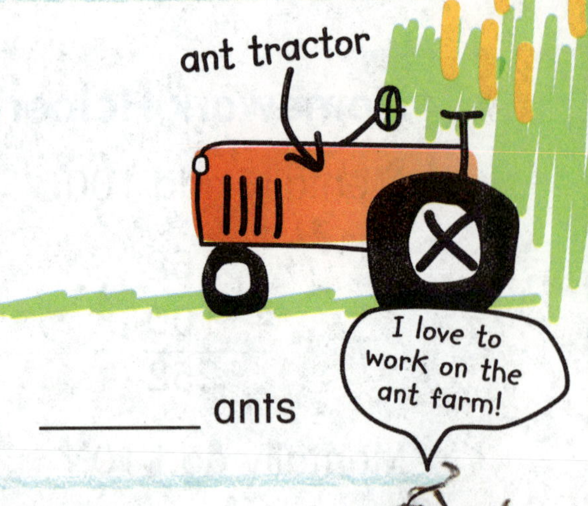

_______ ants

12. Ben has 115 pennies. His dad gives him more pennies. Now Ben has 215 pennies. How many pennies did his dad give him?

_______ pennies

Test Practice

13. Find the sum.

825 + 100 = _______

725	900	925	920
◯	◯	◯	◯

Math at Home Practice saying a number and having your child add 10 or 100 to the number mentally.

Name ....

Check My Progress

Vocabulary Check

Complete each sentence.

tens **regroup** **ones**

1. In the number 352, the 5 is in the __________ place.
2. In the number 352, the 2 is in the __________ place.
3. If the ones add up to more than 9, you must __________.

Concept Check

Make a hundred to add.

4. 25 + 197

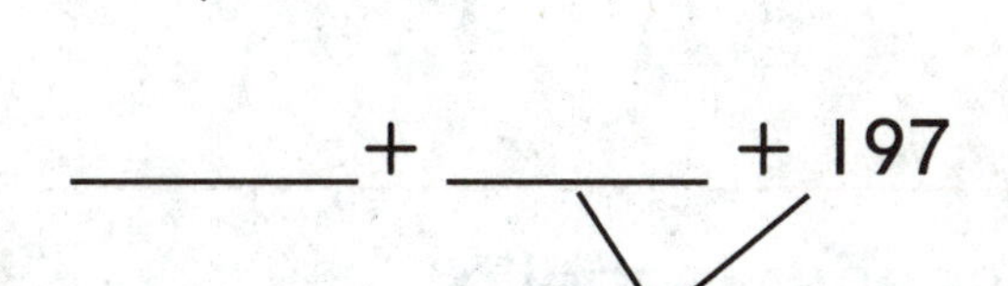

_____ + _____ + 197

22 + _____ = _____

So, 25 + 197 = _____.

5. 314 + 96

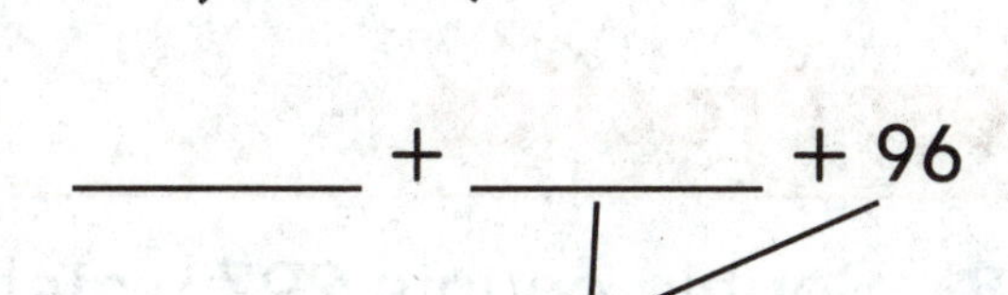

_____ + _____ + 96

310 + _____ = _____

So, 314 + 96 = _____.

Add.

6. $200 + 100 =$ ______

7. $400 + 200 =$ ______

8.
$$\begin{array}{r} 100 \\ +\ 300 \\ \hline \end{array}$$

9.
$$\begin{array}{r} 700 \\ +\ 100 \\ \hline \end{array}$$

10.
$$\begin{array}{r} 500 \\ +\ 400 \\ \hline \end{array}$$

11.
$$\begin{array}{r} 836 \\ +\ 100 \\ \hline \end{array}$$

12.
$$\begin{array}{r} 390 \\ +\ 100 \\ \hline \end{array}$$

13.
$$\begin{array}{r} 567 \\ +\ 10 \\ \hline \end{array}$$

14. $564 + 10 =$ ______

15. $626 + 100 =$ ______

Find the missing addend.

16. $239 +$ ______ $= 249$

17. $853 +$ ______ $= 953$

Test Practice

18. Sophia counts 297 watermelon seeds. Isabella counts 100 more seeds than Sophia. How many seeds does Isabella count?

100	197	297	397
○	○	○	○

Name

Regroup Ones to Add

Lesson 4

ESSENTIAL QUESTION
How can I add three-digit numbers?

Explore and Explain

_____ + _____ = _____ books

Teacher Directions: Model using base-ten blocks. Our library has 146 books about dogs. It has 145 books about cats. How many books about cats and dogs does the library have? Write the number sentence.

See and Show

Find 135 + 328.

Step 1 Add the ones. If there are 10 or more ones, regroup 10 ones as 1 ten. Write the 1 in the tens column.

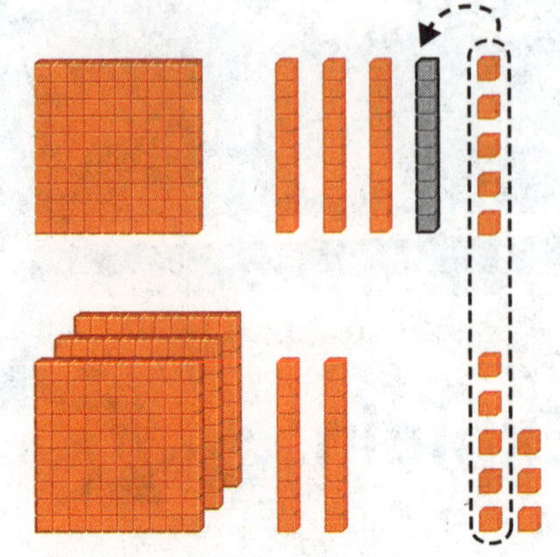

	hundreds	tens	ones
		1	
	1	3	5
+	3	2	8
			3

Step 2 Add the tens.

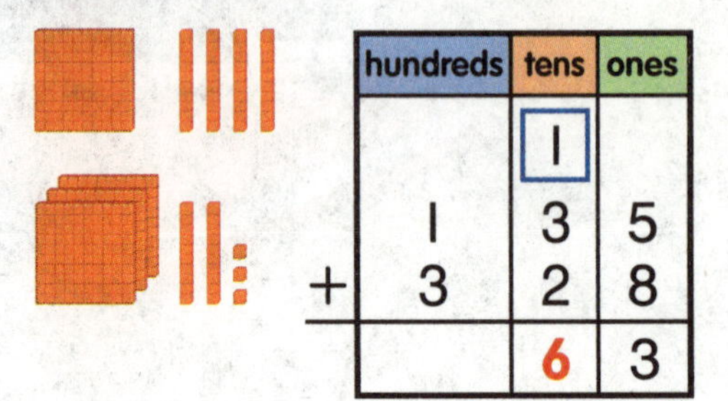

	hundreds	tens	ones
		1	
	1	3	5
+	3	2	8
		6	3

Step 3 Add the hundreds.

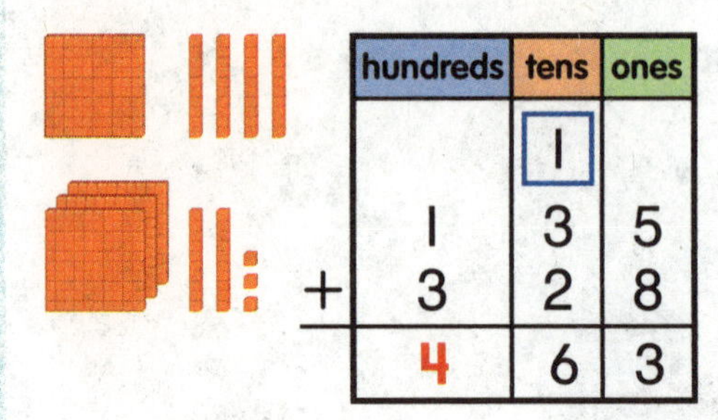

	hundreds	tens	ones
		1	
	1	3	5
+	3	2	8
	4	6	3

So, 135 + 328 = 463.

Use Work Mat 7 and base-ten blocks. Add.

1.

	hundreds	tens	ones
	4	3	6
+	2	4	5

2.

	hundreds	tens	ones
	1	2	7
+	6	4	8

Talk Math How is three-digit addition like two-digit addition?

Name

CCSS

Helpful Hint
Remember to write 1 in the tens column if you regroup.

On My Own

Add. Use base-ten blocks if needed.

3.

	☐	
4	6	8
+	2	3

4.

	☐	
5	6	1
+	2	6

5.

	☐	
4	1	1
+ 3	7	9

6. 236 + 518

7. 468 + 18

8. 427 + 144

9. 306 + 408

10. 28 + 515

11. 273 + 224

12. 749 + 9

13. 146 + 253

14. 135 + 16

15. 305 + 306

16. 607 + 13

17. 782 + 103

Mathematical PRACTICE

Problem Solving

18. There are 247 red cars. There are 438 black cars. How many red and black cars in all?

_______ cars

19. Mr. Archer picked 416 apples last month. He picked 336 apples this month. How many apples did he pick in all?

_______ apples

20. The park ranger counted 105 black bears last year. This year, he counts 128 black bears. How many black bears did he count in all?

_______ black bears

Explain how to regroup ones to make a ten when adding three-digit numbers.

____

____

____

Name ....

My Homework

Lesson 4

Regroup Ones to Add

Homework Helper

Need help? connectED.mcgraw-hill.com

Find 225 + 137.

Step 1 Add the ones. Regroup 10 ones as 1 ten.

Step 2 Add the tens.

Step 3 Add the hundreds.

	hundreds	tens	ones
		1	
	2	2	5
+	1	3	7
	3	6	2

Practice

Add.

1.

	hundreds	tens	ones
		□	
	4	6	3
+	1	1	8

2.

	hundreds	tens	ones
		□	
	1	8	2
+	1	0	9

3.

	hundreds	tens	ones
		□	
	3	3	4
+	1	2	8

4.

		□	
	7	4	4
+	1	3	8

5.

		□	
	3	6	3
+	2	1	9

6.

		□	
	8	2	7
+	1	5	5

Add.

7.	8.	9.
335 + 116	135 + 219	425 + 148

10. 149 children are on a playground. 131 more children come to the playground. How many children are on the playground now?

_____ children

11. There are 152 tadpoles and 138 frogs in a pond. How many frogs and tadpoles are there in all?

_____ frogs and tadpoles

Test Practice

12. Find the sum.

$368 + 619 =$ _____

919 ○ 968 ○ 987 ○ 997 ○

Math at Home Have your child explain how to regroup to add 215 + 215.

Number and Operations in Base Ten
2.NBT.7, 2.NBT.9
CCSS

Name

Regroup Tens to Add

Lesson 5

ESSENTIAL QUESTION
How can I add three-digit numbers?

Explore and Explain

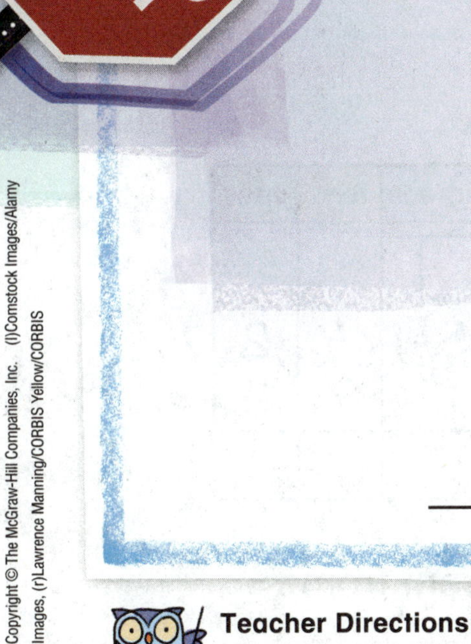

_____ + _____ = _____

Teacher Directions: Model using base-ten blocks. My town has 163 stop signs. There are also 181 street lights. How many stop signs and street lights does my town have? Write the number sentence.

See and Show

Find 375 + 462.

Step 1 Add the ones.

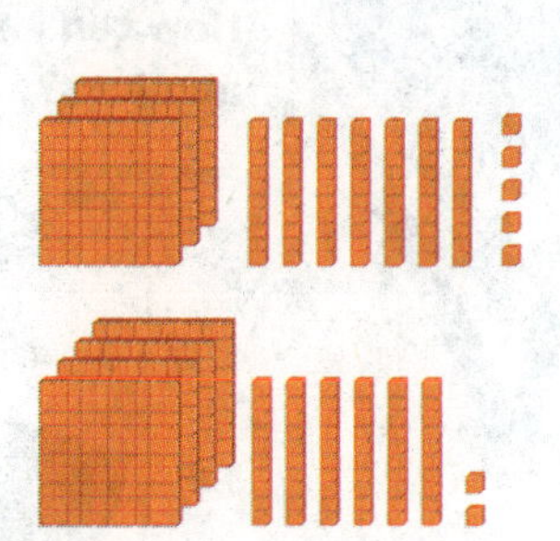

hundreds	tens	ones
3	7	5
+ 4	6	2
		7

Step 2 Add the tens. If there are 10 or more tens, regroup 10 tens as 1 hundred. Write 1 in the hundreds column.

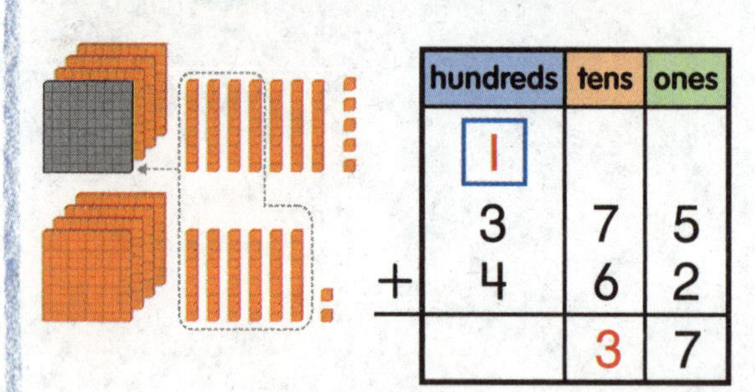

hundreds	tens	ones
1		
3	7	5
+ 4	6	2
	3	7

Step 3 Add the hundreds.

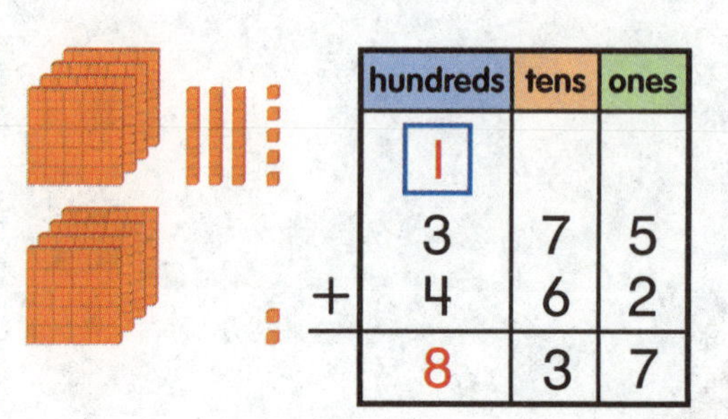

hundreds	tens	ones
1		
3	7	5
+ 4	6	2
8	3	7

So, 375 + 462 = 837.

Use Work Mat 7 and base-ten blocks. Add.

1.

hundreds	tens	ones
☐		
2	4	3
+ 3	8	5

2.

hundreds	tens	ones
☐		
5	6	2
+ 1	7	4

How is regrouping ones different from regrouping tens?

Name

On My Own

Add. Use base-ten blocks if needed.

3.

4	5	6
+ 2	9	1

4.

7	3	2
+ 1	6	7

5.

3	2	4
+	9	3

6.

6	8	4
+	2	4

7.

4	8	5
+ 3	3	2

8.

8	9	5
+	1	1

9. 363 + 281

10. 286 + 121

11. 384 + 134

12. 466 + 60

13. 352 + 475

14. 382 + 51

15. 601 + 281

16. 558 + 360

17. 387 + 122

Problem Solving

18. There are 156 men and 163 women at the town meeting. How many people came to the meeting?

_______ people

19. A forest ranger plants 132 pine trees and 191 oak trees. How many trees did the ranger plant in all?

_______ trees

20. There are 124 red park benches and 185 blue park benches. How many red and blue park benches are there in all?

_______ benches

Write Math How do you know when to regroup?

____

____

____

Name

My Homework

Lesson 5
Regroup Tens to Add

Homework Helper

Need help? connectED.mcgraw-hill.com

Find 353 + 176.

Step 1 Add the ones.

Step 2 Add the tens. If there are 10 or more tens, regroup 10 tens as 1 hundred.

Step 3 Add the hundreds.

So, 353 + 176 = 529.

	hundreds	tens	ones
	1		
	3	5	3
+	1	7	6
	5	2	9

Practice

Add.

1.

	hundreds	tens	ones
	□		
	3	7	5
+	4	5	4

2.

	hundreds	tens	ones
	□		
	2	5	7
+	6	9	1

3.

	hundreds	tens	ones
	□		
	3	2	5
+	1	9	2

4.

	□		
	1	9	8
+	1	2	1

5.

	□		
	6	5	6
+	1	5	3

6.

	□		
	1	8	5
+	2	9	4

Add.

7. $\begin{array}{r} 194 \\ +\ 333 \\ \hline \end{array}$

8. $\begin{array}{r} 352 \\ +\ 281 \\ \hline \end{array}$

9. $\begin{array}{r} 736 \\ +\ 170 \\ \hline \end{array}$

10. Last year there were 252 sunny days. This year there were 164 sunny days. How many sunny days were there in all?

_______ days

11. Logan has 156 baseball cards. Caden has 182 baseball cards. How many baseball cards do they have in all?

_______ cards

Test Practice

12. Find the sum.

$373 + 465 =$ _______

738	838	938	818
○	○	○	○

Math at Home Have your child explain how to regroup tens to solve 185 + 292.

Name ....

Add Three-Digit Numbers

Lesson 6

ESSENTIAL QUESTION
How can I add three-digit numbers?

Explore and Explain

_______ + _______ = _______ people

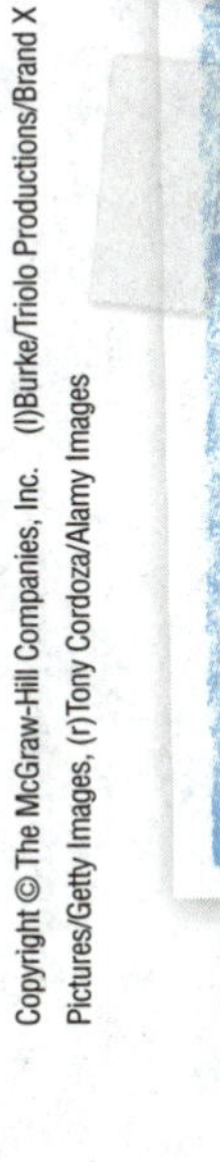

Teacher Directions: Model using base-ten blocks. On Friday, 347 people went to the movies. On Saturday, 485 people went to the movies. How many people went to the movies on these two days in all? Write the number sentence.

See and Show

Find 582 + 268.

Step 1 Add the ones, regroup.

Step 2 Add the tens, regroup.

Step 3 Add the hundreds.

So, 582 + 268 = 850.

	hundreds	tens	ones
	1	1	
	5	8	2
+	2	6	8
	8	5	0

Use Work Mat 7 and base-ten blocks. Add.

1.

	hundreds	tens	ones
	□	□	
	3	5	7
+	4	5	4

2.

	hundreds	tens	ones
	□	□	
	2	6	1
+		6	9

3.

	hundreds	tens	ones
	□	□	
	1	1	8
+	1	9	9

4.

	hundreds	tens	ones
	□	□	
	5	2	3
+	2	8	7

Talk Math Explain how you solved Exercise 2.

Name

CCSS

On My Own

Add. Use base-ten blocks if needed.

5.
□	□	
1	6	6
+ 1	4	7

6.
□	□	
7	2	0
+ 1	9	5

7.
□	□	
4	5	8
+	6	6

8. 34 + 515

9. 632 + 299

10. 327 + 127

11. 129 + 386

12. 103 + 609

13. 123 + 785

14. 843 + 77

15. 422 + 309

16. 215 + 193

17. 395 + 238

18. 725 + 174

19. 662 + 199

Problem Solving

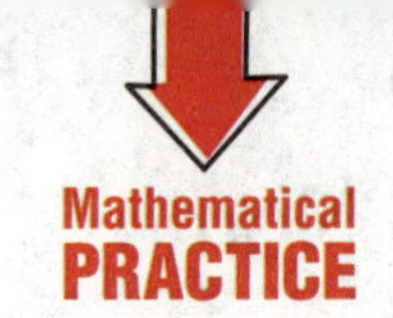

20. One day Keith skipped 145 rocks over the river. The next day he skipped 165 rocks. How many rocks did he skip altogether?

_______ rocks

I get a big KICK out of karate!

21. One karate class has 156 students. The other karate class has 176 students. How many students are in both karate classes?

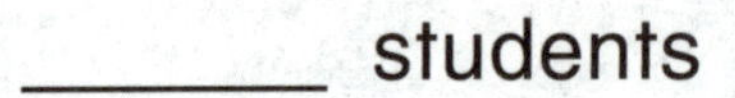

_______ students

22. 389 people went to the zoo on Saturday. 592 people went to the zoo on Sunday. How many people went to the zoo in all?

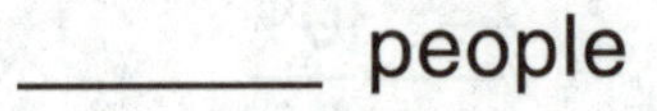

_______ people

Explain how adding three-digit numbers is different than adding two-digit numbers.

____

____

____

Name ________________

My Homework

Lesson 6
Add Three-Digit Numbers

Homework Helper

Need help? connectED.mcgraw-hill.com

Find 536 + 189.

Step 1 There are 10 or more ones, regroup.
Step 2 Add tens. There are 10 or more tens, regroup.
Step 3 Add hundreds.
So, 536 + 189 = 725.

	hundreds	tens	ones
	1	1	
	5	3	6
+	1	8	9
	7	2	5

Practice

Add.

1.

	hundreds	tens	ones
	☐	☐	
	2	5	6
+	4	5	6

2.

	hundreds	tens	ones
	☐	☐	
	4	6	4
+	1	8	5

3.

	hundreds	tens	ones
	☐	☐	
	3	5	9
+	1	6	7

4.

	☐	☐	
	5	7	3
+	3	6	3

5.

	☐	☐	
	3	6	7
+	1	5	5

6.

	☐	☐	
	4	7	7
+	2	3	4

Add.

7. $\begin{array}{r} 285 \\ +\ 229 \\ \hline \end{array}$

8. $\begin{array}{r} 476 \\ +\ 345 \\ \hline \end{array}$

9. $\begin{array}{r} 394 \\ +\ 217 \\ \hline \end{array}$

10. There are 189 animals at one zoo. There are 158 animals at another zoo. How many animals are there at both zoos?

_______ animals

11. Ella has 199 marbles. Kaitlyn has 137 marbles. How many marbles do the girls have in all?

_______ marbles

Test Practice

12. Find the sum.

$369 + 264 =$ _______

633	533	433	133
	○		○

Math at Home Give your child a three-digit addition problem. Have him or her show you how he or she could solve the problem and then write the answer.

Name ______________________________

Rewrite Three-Digit Addition

Lesson 7

ESSENTIAL QUESTION
How can I add three-digit numbers?

Explore and Explain

143 + 167

	hundreds	tens	ones
+			

Be proud of your town!

______ firefighters and police officers

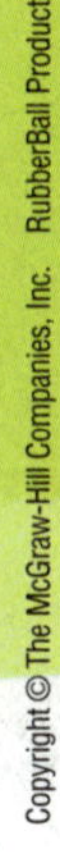

Teacher Directions: There are 143 firefighters and 167 police officers in my town. How many firefighters and police officers are in my town in all? Write the number in the place-value chart and add.

Mathematical PRACTICE

See and Show

You can rewrite a problem to add.

Find 366 + 264.

Step 1 Write one addend below the other addend. Line up the ones, tens, and hundreds.

Step 2 Add. Regroup if necessary.

	hundreds	tens	ones
	☐	☐	
	3	6	6
+	2	6	4
	6	3	0

Rewrite the problems. Add.

1. 634 + 125

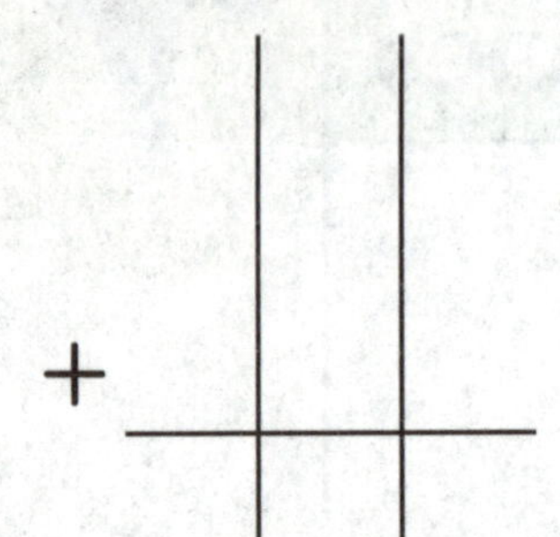

2. 245 + 236

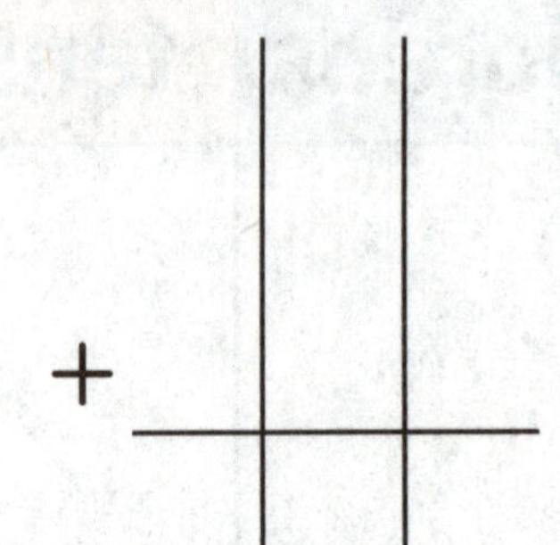

3. 743 + 163

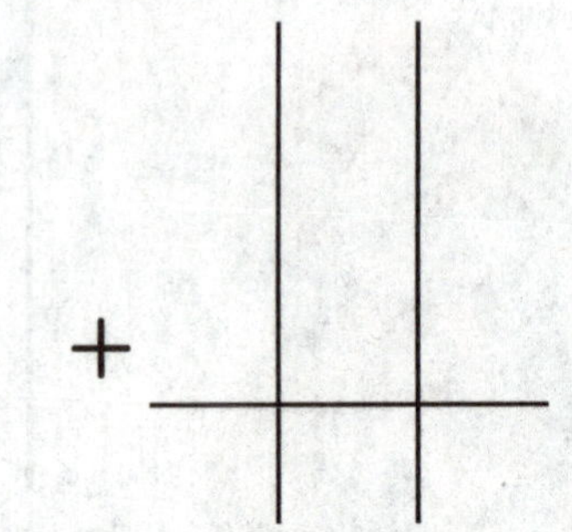

4. 264 + 566

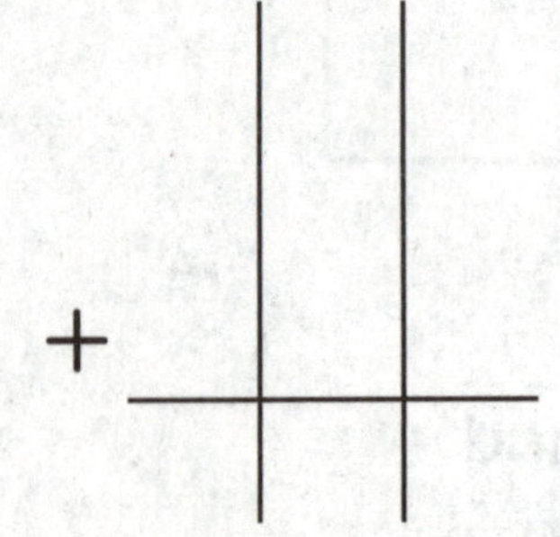

5. 396 + 378

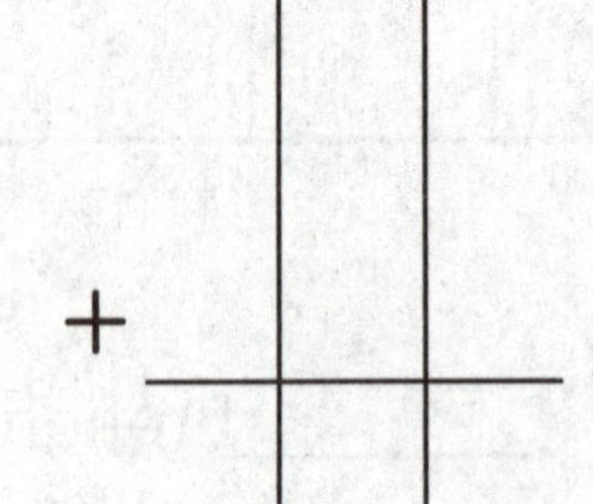

6. 237 + 592

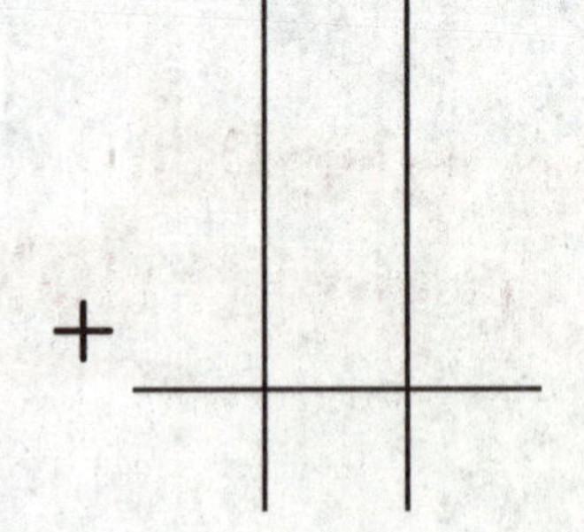

Talk Math How is rewriting three-digit addition different than rewriting two-digit addition?

Name

On My Own

Rewrite the problems. Add.

7. 363 + 278

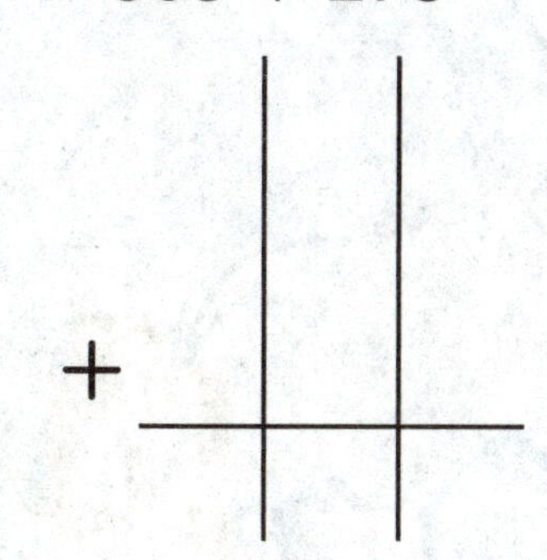

8. 236 + 265

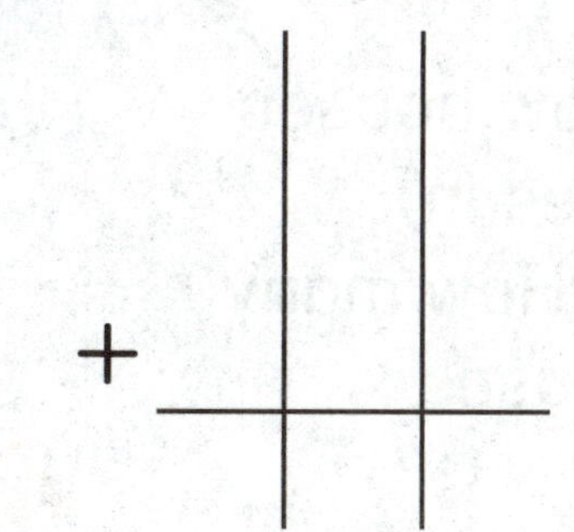

9. 298 + 516

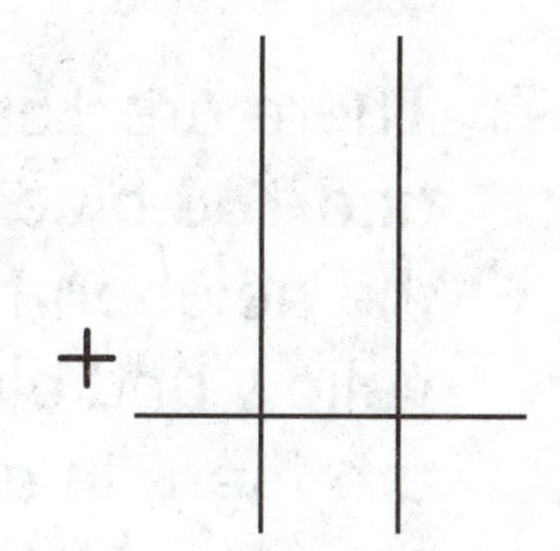

10. 373 + 495

+

11. 195 + 645

+

12. 165 + 454

+

13. 652 + 258

+

14. 268 + 195

+

15. 496 + 349

+

Problem Solving

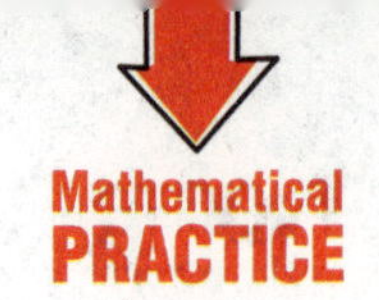

16. 395 ants crawled around the ant hill.
111 more ants came over to the ant hill.
How many ants are there now?

_______ ants

17. There are 153 yellow houses and 168 blue houses in the neighborhood. How many yellow and blue houses are there in all?

_______ yellow and blue houses

18. Last week, Amelia sold 193 cups of lemonade. This week, she sold 129 cups. How many cups of lemonade has she sold in both weeks?

_______ cups

Explain how you rewrite three-digit addition problems to solve.

____

____

____

Name

My Homework

Lesson 7
Rewrite Three-Digit Addition

Homework Helper

Need help? connectED.mcgraw-hill.com

Find 295 + 185.

Step 1 Rewrite.
Step 2 Add. Regroup if necessary.

	1	1	
	2	9	5
+	1	8	5
	4	8	0

Helpful Hint
Line up the ones, tens, and hundreds.

Practice

Rewrite the numbers. Add.

1. 654 + 286

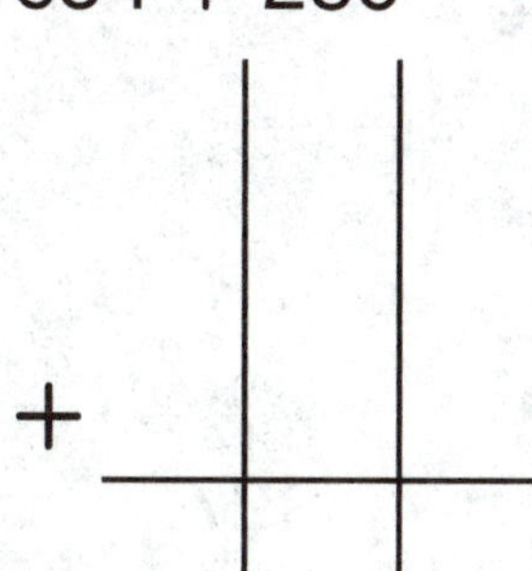

2. 518 + 229

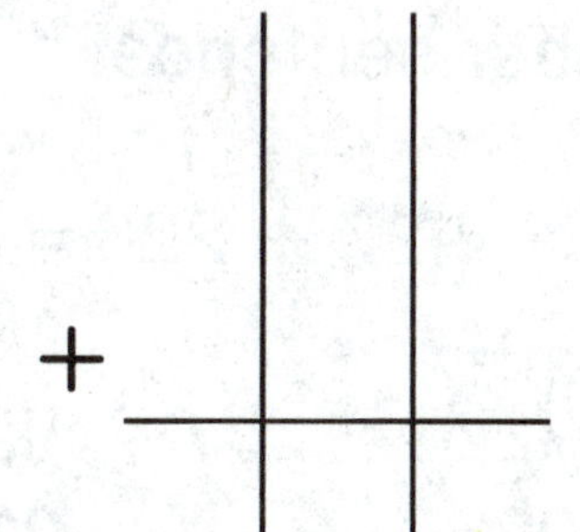

3. 846 + 119

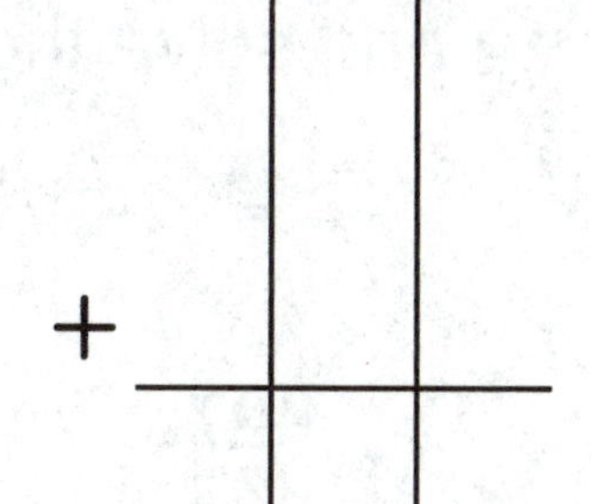

4. 556 + 148

+

5. 685 + 295

+

6. 584 + 371

+

Rewrite the problems. Add.

7. 172 + 217

$$\begin{array}{r} \\ + \\ \hline \end{array}$$

8. 362 + 264

$$\begin{array}{r} \\ + \\ \hline \end{array}$$

9. 239 + 375

$$\begin{array}{r} \\ + \\ \hline \end{array}$$

10. There are 129 boats in the lake. There are 197 jet skis in the lake. How many boats and jet skis are in the lake?

_______ boats and jet skis

Test Practice

11. Mark the answer that shows how to rewrite and solve the number sentence.

427 + 228 = _______

$\begin{array}{r} 427 \\ + 228 \\ \hline 6415 \end{array}$	$\begin{array}{r} ^{3\,11\,17} \\ \not{4}\not{2}\not{7} \\ - 228 \\ \hline 199 \end{array}$	$\begin{array}{r} 427 \\ + 228 \\ \hline 645 \end{array}$	$\begin{array}{r} ^{1} \\ 427 \\ + 228 \\ \hline 655 \end{array}$
○	○	○	○

Math at Home Write 374 + 185 on a piece of paper. Have your child rewrite the problem and solve it.

Name

Problem Solving

STRATEGY: Guess, Check, and Revise

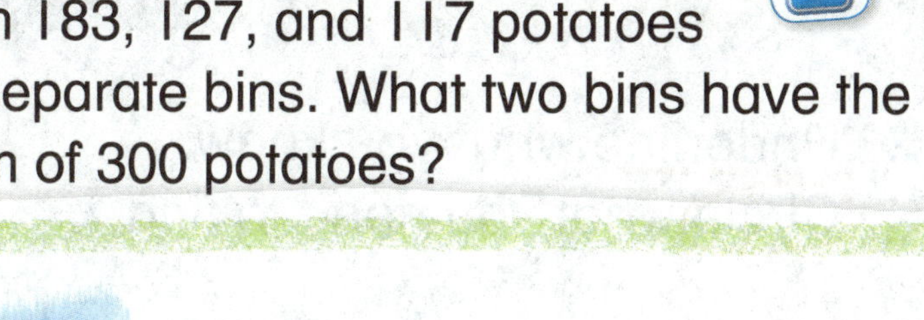

Lesson 8

ESSENTIAL QUESTION
How do I add three-digit numbers?

Watch

A store has three bins of potatoes with 183, 127, and 117 potatoes in separate bins. What two bins have the sum of 300 potatoes?

1 Understand

Underline what you know.
Circle what you need to find.

2 Plan

How will I solve the problem?

3 Solve

Guess, check, and revise.
Guess: 183 and 127
Check: 183 + 127 = 310
Revise: 310 is too high; guess again.
Guess: 183 and 117
Check: 183 + 117 = 300

So, the bins with 183 and 117 have 300 potatoes.

4 Check

Is my answer reasonable? Explain.

Practice the Strategy

A mail carrier delivered 430 letters last month. This month he delivered less letters than last month. He delivered 850 letters in both months in all. How many letters did he deliver this month?

1 Understand Underline what you know.
Circle what you need to find.

2 Plan How will I solve the problem?

3 Solve I will...

4 Check Is my answer reasonable? Explain.

Name

CCSS

Apply the Strategy

1. A store sells raisins in packages of 130, 165, and 170. Rosa wants to buy 300 raisins. Which two packages will she buy?

packages of ______ and ______

2. Alexa's ant farm has 120 ants. She buys another ant farm. She now has 300 ants. How many ants are in the second ant farm?

Guys, look what I found!

______ ants

3. Tao needs to read three books that total 500 pages. Which three books should he read?

Book	Pages
1	115
2	150
3	200
4	185

Books ______, ______, and ______

Review the Strategies

Choose a strategy

- Guess, check, and revise.
- Act it out.
- Use logical reasoning.

4. Clara counted 185 flowers on her walk. Her dad counted 139 flowers. How many flowers did they count altogether?

_____ flowers

5. Chelsea collects sand dollars. The number she has collected is greater than 400 and less than 460. The number in the hundreds place is one less than the number in the tens place. The number in the ones place is two more than the number in the tens place. How many sand dollars has Chelsea collected?

_____ sand dollars

6. One pet store has 289 angel fish. Another pet store has 132 angel fish. How many angel fish do the stores have altogether?

_____ angel fish

Name

My Homework

Lesson 8
Problem Solving: Guess, Check, and Revise

163 girls went to see fireworks. About 300 people went to see the fireworks in all. How many boys went to see the fireworks?

1 Understand Underline what you know.
Circle what you need to find.

2 Plan How will I solve the problem?

3 Solve Guess, check and revise.

Guess: 147 boys
Check: 163 + 147 = 310
Revise: 147 is too high; guess a lower number.
Guess: 137
Check: 163 + 137 = 300
So, 137 boys went to the fireworks.

4 Check Is my answer reasonable?

Underline what you know. Circle what you need to find out. Guess, check, and revise to solve.

1. 350 people went to the movie on Saturday. Fewer people went to the movie on Sunday. About 600 people went to the movies in all. How many people went to the movie on Sunday?

_________ people

2. Pam has 65 purple flowers, 50 yellow flowers, and 75 red flowers in her garden. She needs a total of 125 flowers in 2 colors. Which color of flowers should she pick?

_________________________ flowers

3. My garden has 145 corn stalks. It also has some tomato plants. There are about 245 corn and tomato plants in my garden. About how many tomato plants are in my garden?

_______ tomato plants

Math at Home Tell story problems to your child. Have him or her practice guessing an answer and then checking it.

Name ______________________________

My Review

Chapter 6
Add Three-Digit Numbers

Vocabulary Check

ones hundreds tens regroup place value

Complete each sentence.

1. The numbers 0–9 are ________________.

2. The value given to a digit by its place in a number is its ________________.

3. The numbers 10–99 are ________________.

4. The numbers 100–999 are ________________.

5. When you ________________, you take a number apart and write it in a new way.

Concept Check

Add.

6. $\begin{array}{r} 200 \\ +\ 100 \\ \hline \end{array}$

7. $\begin{array}{r} 400 \\ +\ 500 \\ \hline \end{array}$

8. $\begin{array}{r} 500 \\ +\ 300 \\ \hline \end{array}$

Add.

9. 354 + 100

10. 327 + 10

11. 356 + 100

12. 275 + 116

13. 434 + 247

14. 367 + 226

15. 777 + 133

16. 397 + 159

17. 487 + 247

Rewrite the problems. Add.

18. 395 + 154

+

19. 225 + 353

+

20. 348 + 273

+

CCSS

Name ....

Problem Solving

21. Nathan collected 109 pine cones. Elizabeth collected 100 pine cones. How many pine cones did they collect in all?

_______ pine cones

22. Olivia bakes 154 cookies with her grandma. Connor bakes 146 cookies with his grandma. How many cookies do Olivia and Connor bake in all?

_______ cookies

Test Practice

23. Last month, Riley watched 115 hours of television. This month, she will watch fewer hours of television. She wants to watch 200 hours for both months. How many hours can Riley watch this month?

85	95	185	315
○	○	○	○

Reflect

Chapter 6

Answering the Essential Question

Show the ways you can add three-digit numbers.

ESSENTIAL QUESTION

How do I add three-digit numbers?

Regroup ones.

$$\begin{array}{r} 101 \\ +\ 129 \\ \hline \end{array}$$

Regroup tens.

$$\begin{array}{r} 234 \\ +\ 383 \\ \hline \end{array}$$

Regroup tens and ones.

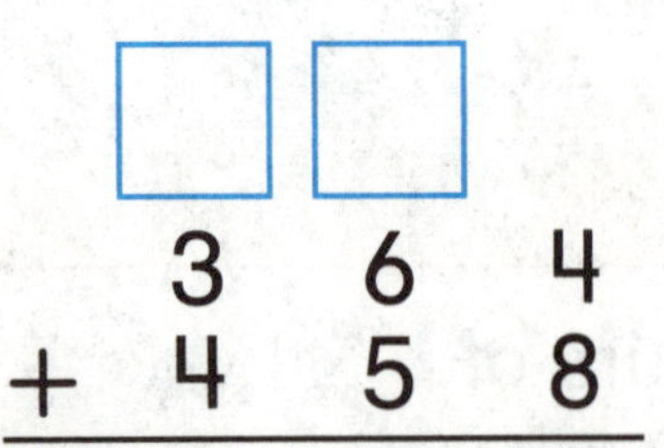

$$\begin{array}{r} 364 \\ +\ 458 \\ \hline \end{array}$$

Add mentally.

$358 + 100 = \_\_\_\_$

$782 + \_\_\_\_ = 792$

You will go far!

Chapter 7

Subtract Three-Digit Numbers

ESSENTIAL QUESTION

How can I subtract three-digit numbers?

My School is Cool!

Watch a video!

My Common Core State Standards

Number and Operations in Base Ten

2.NBT.7 Add and subtract within 1000, using concrete models or drawings and strategies based on place value, properties of operations, and/or the relationship between addition and subtraction; relate the strategy to a written method. Understand that in adding or subtracting three-digit numbers, one adds or subtracts hundreds and hundreds, tens and tens, ones and ones; and sometimes it is necessary to compose or decompose tens or hundreds.

2.NBT.8 Mentally add 10 or 100 to a given number 100–900, and mentally subtract 10 or 100 from a given number 100–900.

2.NBT.9 Explain why addition and subtraction strategies work, using place value and the properties of operations.

Standards for Mathematical PRACTICE

1. Make sense of problems and persevere in solving them.
2. Reason abstractly and quantitatively.
3. Construct viable arguments and critique the reasoning of others.
4. Model with mathematics.
5. Use appropriate tools strategically.
6. Attend to precision.
7. Look for and make use of structure.
8. Look for and express regularity in repeated reasoning.

= focused on in this chapter

Name ....

Am I Ready?

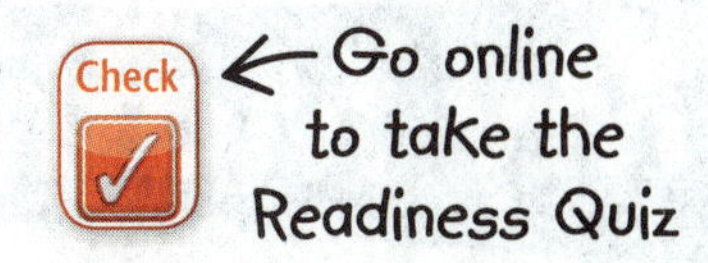

Subtract.

1. $\begin{array}{r} 9 \\ -\ 6 \\ \hline \end{array}$

2. $\begin{array}{r} 8 \\ -\ 8 \\ \hline \end{array}$

3. $\begin{array}{r} 17 \\ -\ 8 \\ \hline \end{array}$

4. $\begin{array}{r} 14 \\ -\ 5 \\ \hline \end{array}$

5. $\begin{array}{r} 60 \\ -\ 30 \\ \hline \end{array}$

6. $\begin{array}{r} 87 \\ -\ 53 \\ \hline \end{array}$

7. $\begin{array}{r} 62 \\ -\ 37 \\ \hline \end{array}$

8. $\begin{array}{r} 72 \\ -\ 25 \\ \hline \end{array}$

9. $\begin{array}{r} 95 \\ -\ 14 \\ \hline \end{array}$

10. $\begin{array}{r} 90 \\ -\ 60 \\ \hline \end{array}$

11. $\begin{array}{r} 88 \\ -\ 61 \\ \hline \end{array}$

12. $\begin{array}{r} 43 \\ -\ 25 \\ \hline \end{array}$

13. Aiden sees eight squirrels. Two run away. How many squirrels does Aiden see now?

_____ squirrels

How Did I Do? **Shade the boxes to show the problems you answered correctly.**

1	2	3	4	5	6	7	8	9	10	11	12	13

Name ....

My Math Words

Review Vocabulary

difference ones regroup subtract tens

Write a review word that has the same meaning as the underlined words or numbers.

I will take away numbers.

____

57 – 39

____

I will take apart the ones.

____

The answer is 18.

____

23 – 12

____

The answer is 11.

____

My Vocabulary Cards

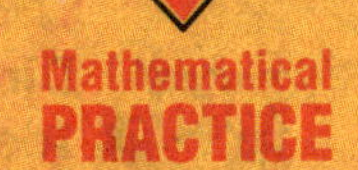

Directions:
Ideas for Use

- Ask students to use the blank cards to draw or write words that will help them with concepts like three-digit subtraction or subtract across zeros.
- Have students use the blank cards to write basic subtraction facts. They should write the answer on the back of each card.

FOLDABLES® Follow the steps on the back to make your Foldable.

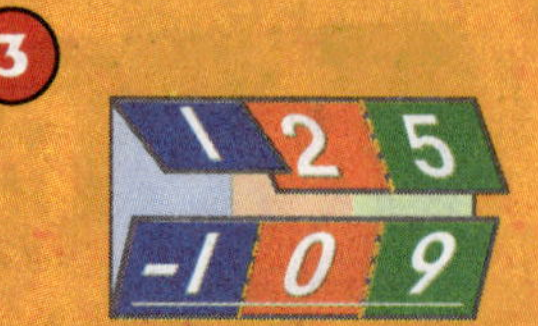

1 2 5

-1 0 9

Name ....

Take Apart Hundreds to Subtract

Lesson 1

ESSENTIAL QUESTION
How can I subtract three-digit numbers?

Explore and Explain

Stick together. We don't want to get lost!

_______ crayons

Teacher Directions: At the beginning of the year, Mrs. Jones put 403 new crayons in the crayon bucket. After the first month of school, 97 crayons were lost. How many crayons were left in the bucket? Use base-ten blocks to help you solve the problem. Draw the blocks you used and write the answer.

See and Show

You can take apart hundreds to subtract mentally.

Take apart 800 as 700 and 100 since it is easier to subtract 98 from 100.

800 − 98

700 100

100 − 98 = 2

700 + 2 = 702

So, 800 − 98 = 702.

Take apart 401 as 301 and 100 since it is easier to subtract 97 from 100.

401 − 97

301 100

100 − 97 = 3

301 + 3 = 304

So, 401 − 97 = 304.

Take apart hundreds to subtract.

1. 300 − 99

_____ _____

_____ − 99 = _____

_____ + _____ = _____

So, 300 − 99 = _____.

2. 422 − 98

_____ _____

_____ − 98 = _____

_____ + _____ = _____

So, 422 − 98 = _____.

Talk Math In Exercise 2, why is 2 added back?

Name ______________________

On My Own

Take apart hundreds to subtract.

3. 240 − 99

______ ______

______ − 99 = ______

______ + ______ = ______

So, 240 − 99 = ______.

4. 700 − 98

______ ______

______ − 98 = ______

______ + ______ = ______

So, 700 − 98 = ______.

5. 542 − 97

______ ______

______ − 97 = ______

______ + ______ = ______

So, 542 − 97 = ______.

6. 702 − 98

______ ______

______ − 98 = ______

______ + ______ = ______

So, 702 − 98 = ______.

7. 200 − 97

______ ______

______ − 97 = ______

______ + ______ = ______

So, 200 − 97 = ______.

8. 711 − 99

______ ______

______ − 99 = ______

______ + ______ = ______

So, 711 − 99 = ______.

Problem Solving

Mathematical PRACTICE

9. Eli counts 335 sunflower seeds. He places 98 of the seeds in a bowl. How many seeds were left?

_____ sunflower seeds

10. Rachel has 148 blueberries. She gives 97 blueberries to her brother. How many blueberries does Rachel have left?

_____ blueberries

11. 382 people were in our school auditorium. 99 people left. How many people are still in the auditorium?

_____ people

HOT Problem Explain in words how to solve 700 − 97.

____

____

____

Name ....

My Homework

Lesson 1
Take Apart Hundreds to Subtract

Homework Helper

Need help? connectED.mcgraw-hill.com

You can take apart hundreds to mentally subtract 324 – 99.

Take apart 324 as 224 and 100.

It is easier to subtract 99 from 100.

324 – 99

224 100

100 – 99 = 1

224 + 1 = 225

So, 324 – 99 = 225.

Practice

Take apart hundreds to subtract.

1. 835 – 98

______ ______

______ – ______ = ______

______ + ______ = ______

So, 835 – 98 = ______.

2. 748 – 97

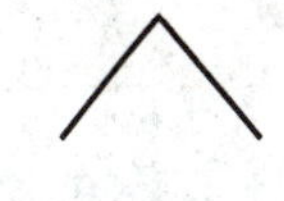

______ ______

______ – ______ = ______

______ + ______ = ______

So, 748 – 97 = ______.

Take apart hundreds to subtract.

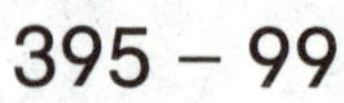

3. 395 − 99

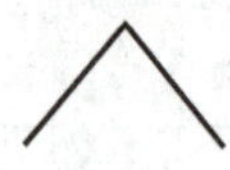

______ ______

______ − ______ = ______

______ + ______ = ______

So, 395 − 99 = ______.

4. 600 − 97

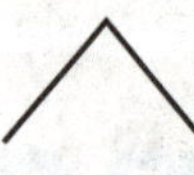

______ ______

______ − ______ = ______

______ + ______ = ______

So, 600 − 97 = ______.

5. Students go to school for 180 days. 96 school days were sunny. How many school days were not sunny?

______ days

Test Practice

6. How would you take apart 355 to solve 355 − 94?

355 and 100

300 and 55

255 and 100

255 and 94

Math at Home Have your child solve 321 − 99 by taking apart hundreds to subtract.

Name

Subtract Hundreds

Lesson 2

ESSENTIAL QUESTION
How can I subtract three-digit numbers?

Explore and Explain

_______ steps

Teacher Directions: There are 400 steps in my school. My class decided to walk up all of the steps. We have walked up 300 steps already. How many steps do we have left? Use base-ten blocks. Draw the blocks. Write how many steps are left.

See and Show

You can use basic subtraction facts to help subtract hundreds.

Find 500 − 300.

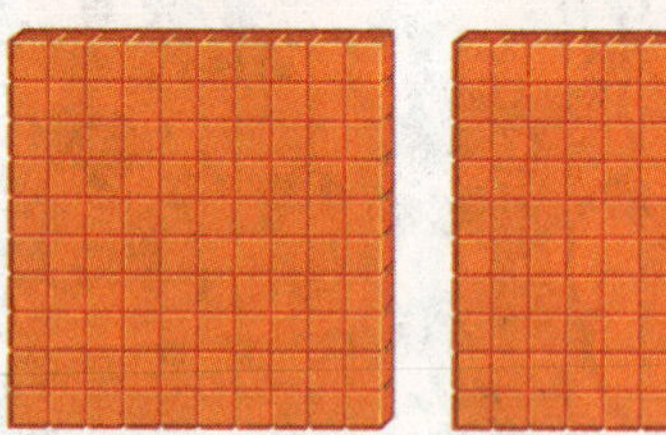

Helpful Hint
You know 5 − 3 = 2.
So, 500 − 300 = 200.

	5	hundreds
−	3	hundreds
	2	hundreds

500 − 300 = 200

Subtract.

1. 800 − 100 = ______
2. 200 − 200 = ______
3. 700 − 100 = ______
4. 400 − 200 = ______
5. 500 − 100 = ______
6. 700 − 300 = ______

Talk Math What subtraction fact can you use to find 900 − 800?

Name ....

Helpful Hint
Use the subtraction facts you know to subtract hundreds.

On My Own

Subtract.

7. 800 − 300 = ______

8. 600 − 600 = ______

9. 300 − 100 = ______

10. 900 − 200 = ______

11. 800 − 400 = ______

12. 700 − 500 = ______

13. 200 − 200 = ______

14. 400 − 100 = ______

15. $\begin{array}{r} 700 \\ -\ 400 \\ \hline \end{array}$

16. $\begin{array}{r} 800 \\ -\ 200 \\ \hline \end{array}$

17. $\begin{array}{r} 300 \\ -\ 300 \\ \hline \end{array}$

18. $\begin{array}{r} 500 \\ -\ 400 \\ \hline \end{array}$

19. $\begin{array}{r} 500 \\ -\ 300 \\ \hline \end{array}$

20. $\begin{array}{r} 200 \\ -\ 100 \\ \hline \end{array}$

21. $\begin{array}{r} 600 \\ -\ 200 \\ \hline \end{array}$

22. $\begin{array}{r} 900 \\ -\ 700 \\ \hline \end{array}$

23. $\begin{array}{r} 700 \\ -\ 200 \\ \hline \end{array}$

24. $\begin{array}{r} 400 \\ -\ 300 \\ \hline \end{array}$

25. $\begin{array}{r} 800 \\ -\ 0 \\ \hline \end{array}$

26. $\begin{array}{r} 800 \\ -\ 800 \\ \hline \end{array}$

Problem Solving

27. 700 people came to my school's spring concert. 500 people came to our winter concert. How many more people came to the spring concert?

_______ people

28. My school made flags for Memorial Day. There were 600 flags. 200 students took their flags home. How many flags are left?

Stars and stripes FOREVER!

_______ flags

29. The floor in my classroom has 200 tiles. 100 tiles are blue. The rest of the tiles are white. How many white tiles are there?

_______ tiles

How is subtracting hundreds like subtracting ones?

____

____

____

Name

My Homework

Lesson 2
Subtract Hundreds

Homework Helper

Need help? connectED.mcgraw-hill.com

Find 300 − 200.

Helpful Hint
You know 3 - 2 = 1.
So, 300 - 200 = 100.

3 hundreds
− 2 hundreds
1 hundred

300 − 200 = 100

Practice

Subtract.

1. 700 − 300 = ______

2. 500 − 200 = ______

3. 300 − 100 = ______

4. 100 − 100 = ______

5. 600 − 400 = ______

6. 800 − 700 = ______

7. 900 − 100 = ______

8. 500 − 100 = ______

9. 400 − 400 = ______

10. 400 − 200 = ______

Solve each word problem.

11. 800 people went to watch the ballet. 200 people left early. How many people stayed at the ballet?

_______ people

12. 500 boys and girls went to the basketball game. 300 girls were at the game. How many boys were at the game?

_______ boys

13. The cafeteria had pizza slices and hot dogs for lunch. 900 lunches were sold. 300 people bought hot dogs. How many pizza slices were sold?

_______ pizza slices

Test Practice

14. Which number sentence could help you solve $700 - 500$?

$7 - 2$	$2 + 5$	$5 + 2$	$7 - 5$
○	○	○	○

Math at Home Ask your child what number is 100 less than 500.

Name

Mentally Subtract 10 or 100

Lesson 3

ESSENTIAL QUESTION
How can I subtract three-digit numbers?

Explore and Explain

_______ students

Teacher Directions: 135 students are on the playground. 100 students go inside. How many students are left on the playground? Use base-ten blocks to solve. Write the number.

See and Show

To subtract 100 or 10, think of facts you know.

Mentally subtract 100.

$$\begin{array}{r} 422 \\ -\ 100 \\ \hline \end{array}$$

$$\begin{array}{r} 422 \\ -\ 100 \\ \hline 322 \end{array}$$

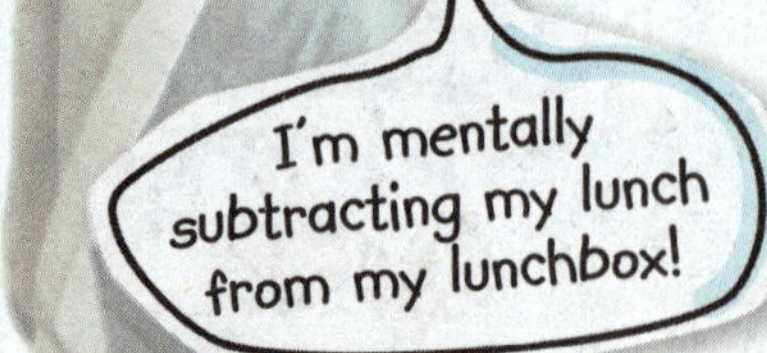

Mentally subtract 10.

$$\begin{array}{r} 460 \\ -\ 10 \\ \hline \end{array}$$

$$\begin{array}{r} 460 \\ -\ 10 \\ \hline 450 \end{array}$$

Subtract.

1. $\begin{array}{r} 329 \\ -100 \\ \hline \end{array}$

2. $\begin{array}{r} 820 \\ -\ 10 \\ \hline \end{array}$

3. $\begin{array}{r} 363 \\ -100 \\ \hline \end{array}$

4. $\begin{array}{r} 678 \\ -\ 10 \\ \hline \end{array}$

5. $\begin{array}{r} 724 \\ -\ 10 \\ \hline \end{array}$

6. $\begin{array}{r} 164 \\ -\ 10 \\ \hline \end{array}$

7. $900 - 100 =$ ______

8. $743 - 10 =$ ______

Talk Math Tell how to mentally subtract 10 or 100.

Name ....

CCSS

On My Own

Subtract.

9. $491 - 100 =$ ____

10. $942 - 10 =$ ____

11. $770 - 10 =$ ____

12. $672 - 100 =$ ____

13. $853 - 100 =$ ____

14. $269 - 10 =$ ____

15. $368 - 10 =$ ____

16. $374 - 100 =$ ____

17. $982 - 100 =$ ____

18. 498 − 100 = ____

19. 533 − 100 = ____

Find the missing number.

20. 434 − ____ = 424

21. 371 − ____ = 271

22. 738 − ____ = 638

23. 270 − ____ = 260

Problem Solving

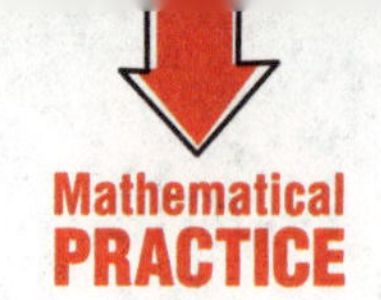

24. 863 flowers bloomed on the school playground. 100 of the flowers were eaten by bugs. How many flowers are on the playground now?

_________ flowers

25. 299 people came to the school play. 10 people had to leave early. How many people were still at the play?

_________ people

26. 423 wildflowers grew in the field. Embry picked 10 of them. How many wildflowers are left in the field?

_________ wildflowers

Explain how you would mentally subtract 10 from 900.

____

____

____

Name ....

My Homework

Lesson 3
Mentally Subtract 10 or 100

Homework Helper

Need help? connectED.mcgraw-hill.com

Mentally subtract 100.

$$\begin{array}{r} 567 \\ -\ 100 \\ \hline 467 \end{array}$$

Helpful Hint
You know that
5 - 1 = 4.

Mentally subtract 10.

$$\begin{array}{r} 567 \\ -\ 10 \\ \hline 557 \end{array}$$

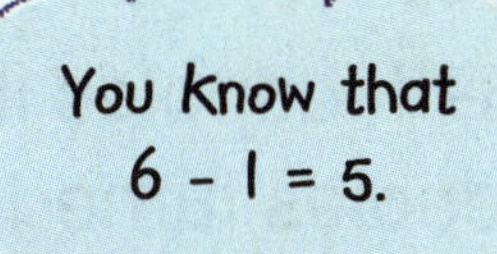

Practice

Subtract.

1. $\begin{array}{r} 477 \\ -\ 100 \\ \hline \end{array}$

2. $\begin{array}{r} 378 \\ -\ 10 \\ \hline \end{array}$

3. $\begin{array}{r} 879 \\ -\ 100 \\ \hline \end{array}$

4. $\begin{array}{r} 245 \\ -\ 10 \\ \hline \end{array}$

5. $\begin{array}{r} 849 \\ -\ 100 \\ \hline \end{array}$

6. $\begin{array}{r} 320 \\ -\ 10 \\ \hline \end{array}$

Find the missing numbers.

7. 358 − ______ = 258

8. 843 − ______ = 833

9. 954 − ______ = 944

10. 700 − ______ = 600

11. Joan counts 143 birds by the pond. 100 birds fly away. How many birds are left by the pond?

______ birds

12. There were 694 tadpoles in the lake. 100 of the tadpoles turned into frogs. How many tadpoles were left in the lake?

______ tadpoles

Test Practice

13. Find 363 − 100.

163 ○ 263 ○ 363 ○ 463 ○

Math at Home Practice saying a number and having your child subtract 10 or 100 from the number mentally.

Name ______________________

Check My Progress

Vocabulary Check

Draw lines to match.

1.	**difference**	To find the difference between two sets.
2.	**subtract**	To take apart a number to write it in a new way.
3.	**regroup**	The answer to a subtraction problem.

Concept Check

Take apart addends to subtract.

4. 345 − 98

______ ______

______ − ______ = ______

______ + ______ = ______

So, 345 − 98 = ______.

5. 926 − 99

______ ______

______ − ______ = ______

______ + ______ = ______

So, 926 − 99 = ______.

Subtract.

6. 900 − 400 = ______

7. 700 − 300 = ______

8. $\begin{array}{r} 393 \\ -\ 100 \\ \hline \end{array}$

9. $\begin{array}{r} 264 \\ -\ 10 \\ \hline \end{array}$

10. $\begin{array}{r} 737 \\ -\ 100 \\ \hline \end{array}$

Find the missing number.

11. 394 − ______ = 294

12. 842 − ______ = 832

13. 535 − ______ = 435

14. 253 − ______ = 243

15. Brooklyn read 125 books over the summer.
Aubrey read 10 less books than Brooklyn.
How many books did Aubrey read?

______ books

Test Practice

16. **Find 300 − 300.**

600 ○ 300 ○ 100 ○ 0 ○

Name

Regroup Tens

Lesson 4

ESSENTIAL QUESTION
How can I subtract three-digit numbers?

Explore and Explain

Only 360 more and I break the school record!

Go!

_______ jump ropes

Teacher Directions: Use base-ten blocks to solve. Mr. Hicks, our gym teacher, had 145 jump ropes. 36 jump ropes broke this year. How many jump ropes does Mr. Hicks have left? Write the number.

See and Show

Find 652 − 429.

Step 1 Subtract the ones. You cannot subtract 9 from 2. Regroup 1 ten as 10 ones.

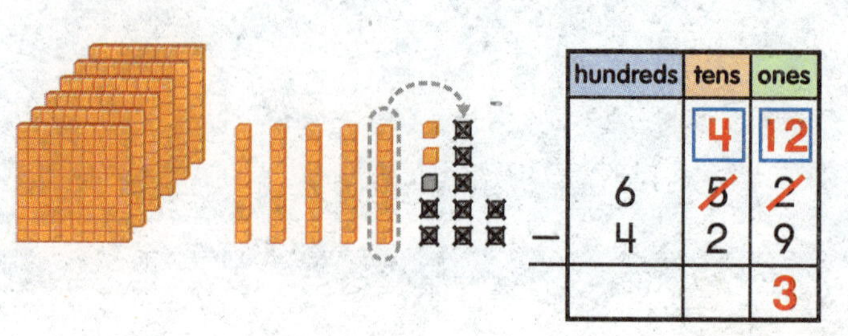

Step 2 Subtract the tens.

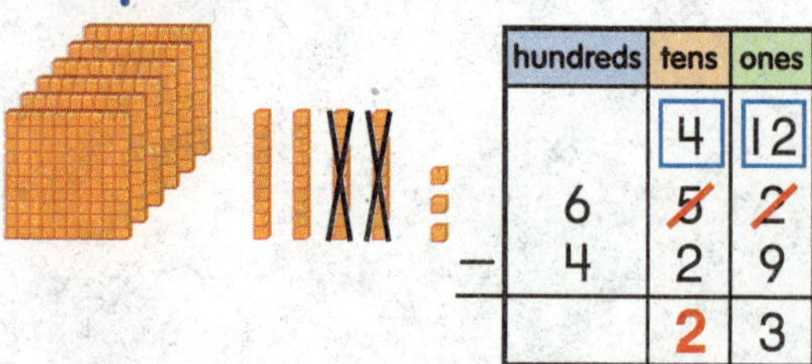

Step 3 Subtract the hundreds.

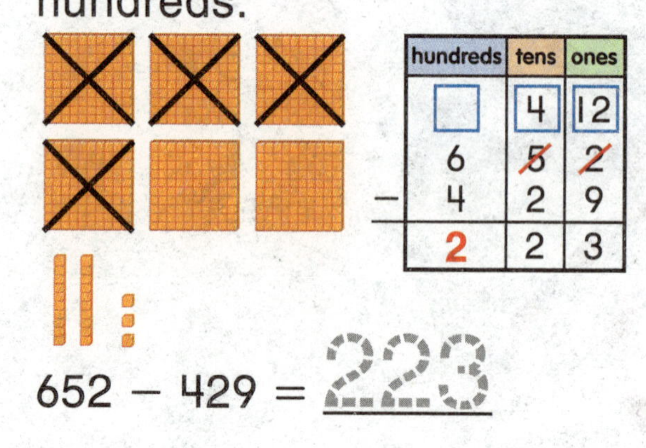

652 − 429 = 223

Use Work Mat 7 and base-ten blocks. Subtract.

1.

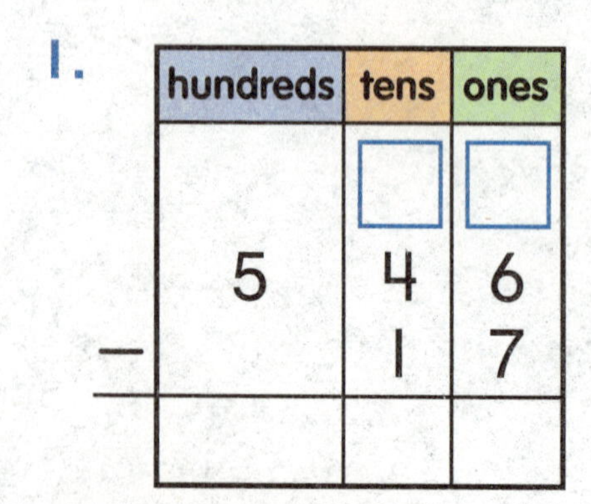

2.

	hundreds	tens	ones
		□	□
	7	8	3
−	4	3	9

Talk Math How is subtracting three-digit numbers like subtracting two-digit numbers?

Name

On My Own

Use Work Mat 7 and base-ten blocks. Subtract.

3.

	hundreds	tens	ones
		□	□
	3	8	2
−	1	2	8

4.

	hundreds	tens	ones
		□	□
	4	6	7
−		4	9

5.

		□	□
	5	7	5
−		6	6

6.

		□	□
	8	6	3
−	2	1	8

7.

		□	□
	2	6	4
−	1	3	5

8. $\begin{array}{r} 754 \\ -\ 507 \\ \hline \end{array}$

9. $\begin{array}{r} 455 \\ -\ 326 \\ \hline \end{array}$

10. $\begin{array}{r} 930 \\ -\ 428 \\ \hline \end{array}$

11. $\begin{array}{r} 780 \\ -\ 436 \\ \hline \end{array}$

12. $\begin{array}{r} 652 \\ -\ 35 \\ \hline \end{array}$

13. $\begin{array}{r} 931 \\ -\ 6 \\ \hline \end{array}$

14. $\begin{array}{r} 387 \\ -\ 18 \\ \hline \end{array}$

15. $\begin{array}{r} 423 \\ -\ 119 \\ \hline \end{array}$

16. $\begin{array}{r} 540 \\ -\ 15 \\ \hline \end{array}$

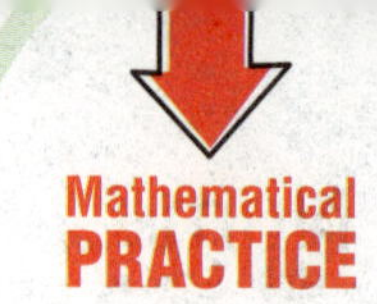

Problem Solving

17. Ada jumped 382 times in a row. Helen jumped 277 times in a row. How many more jumps in a row did Ada do than Helen?

_______ jumps

18. A store had 472 bouncy balls. They sold 155. How many bouncy balls were left?

_______ bouncy balls

19. Lisa hit 294 golf balls in one day. She hit 149 in the morning. How many golf balls did she hit in the afternoon?

_______ golf balls

HOT Problem Andre wrote 381 − 165 = 224. Tell why Andre is wrong. Make it right.

____

____

____

Name ....

My Homework

Lesson 4
Regroup Tens

Homework Helper

Need help? connectED.mcgraw-hill.com

Find 362 − 145.

Step 1 Subtract the ones. You cannot subtract 5 from 2. Regroup if needed.

Step 2 Subtract the tens.

Step 3 Subtract the hundreds.

	hundreds	tens	ones
		5	12
	3	~~6~~	~~2~~
−	1	4	5
	2	1	7

Practice

Subtract.

1.

	hundreds	tens	ones
		☐	☐
	8	3	5
−	2	1	6

2.

	hundreds	tens	ones
		☐	☐
	9	5	2
−	6	3	7

3.

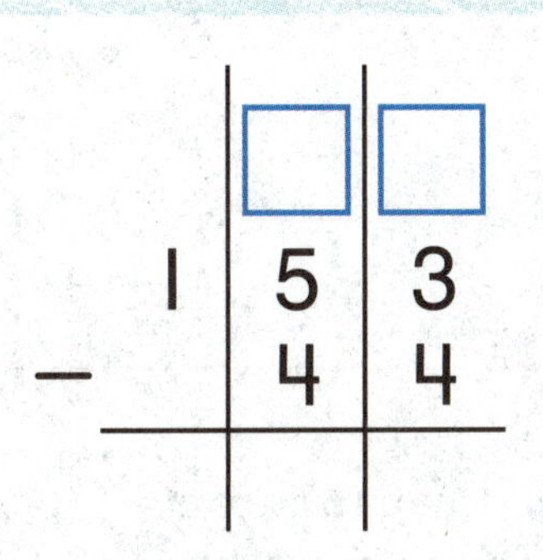

		☐	☐
	1	5	3
−		4	4

4.

		☐	☐
	6	4	2
−	2	1	8

5.

		☐	☐
	7	3	4
−	2	2	8

Subtract.

6. 153 − 47	**7.** 642 − 215	**8.** 754 − 225	**9.** 438 − 129
10. 362 − 148	**11.** 762 − 349	**12.** 647 − 518	**13.** 377 − 163

14. 725 people came to my school's pancake breakfast. 318 of those people were children. How many adults were at the breakfast?

_______ adults

Test Practice

15. Which subtraction problem needs regrouping to solve?

392 − 222 ○ 385 − 266 ○

692 − 321  295 − 172 ○

Math at Home Write a subtraction problem that requires regrouping tens on a piece of paper for your child to solve. Have your child explain each step to solving the problem. Try one-, two-, and three-digit subtraction problems.

Name

Regroup Hundreds

Lesson 5

ESSENTIAL QUESTION
How can I subtract three-digit numbers?

Explore and Explain

I am the most famous macaroni artist!

_____ pieces of macaroni

Teacher Directions: Use base-ten blocks to solve. Gwen put 214 pieces of macaroni on her picture in art class. 120 pieces fell off. How many pieces of macaroni are still on the picture? Draw the blocks you used. Write the number.

See and Show

Find 539 − 285.

Step 1

Subtract the ones.

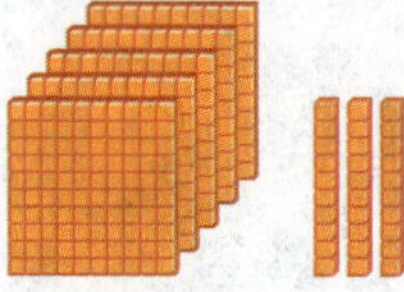

	hundreds	tens	ones
	5	3	9
−	2	8	5
			4

Step 2

Subtract the tens. You cannot subtract 8 from 3. Regroup 1 hundred as 10 tens.

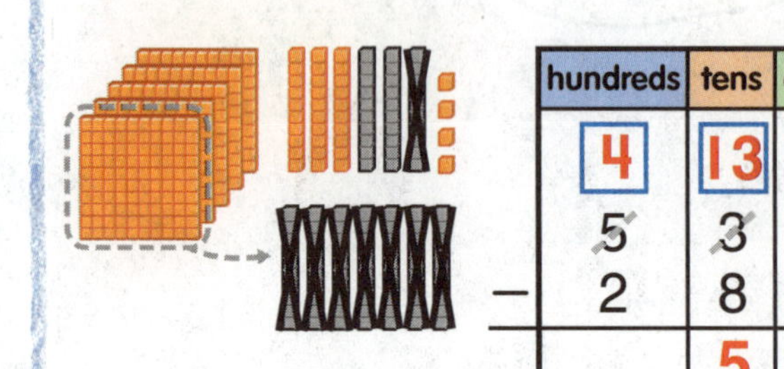

	hundreds	tens	ones
	4	13	
	~~5~~	~~3~~	9
−	2	8	5
		5	4

Step 3

Subtract the hundreds.

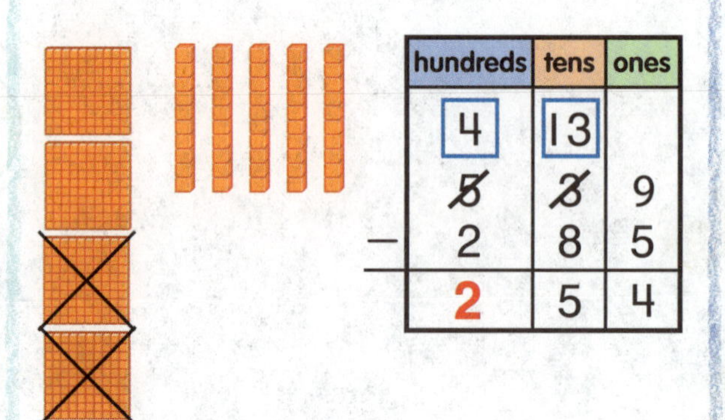

	hundreds	tens	ones
	4	13	
	~~5~~	~~3~~	9
−	2	8	5
	2	5	4

So, 539 − 285 = 254

Use Work Mat 7 and base-ten blocks. Subtract.

1.

	hundreds	tens	ones
	☐	☐	
	5	6	8
−	1	9	7

2.

	hundreds	tens	ones
	☐	☐	
	6	2	8
−	4	4	2

Talk Math How do you know when to regroup?

Name

On My Own

Use Work Mat 7 and base-ten blocks. Subtract.

3.

	hundreds	tens	ones
	☐	☐	
	5	8	6
−	2	9	5

4.

	hundreds	tens	ones
	☐	☐	
	6	3	8
−		4	3

5.

	☐	☐	
	4	5	9
−		6	9

6.

	☐	☐	
	7	3	9
−	5	4	1

7.

	☐	☐	
	8	2	7
−	2	4	7

8. $\begin{array}{r} 638 \\ -\ 36 \\ \hline \end{array}$

9. $\begin{array}{r} 232 \\ -\ 170 \\ \hline \end{array}$

10. $\begin{array}{r} 948 \\ -\ 472 \\ \hline \end{array}$

11. $\begin{array}{r} 565 \\ -\ 272 \\ \hline \end{array}$

12. $\begin{array}{r} 640 \\ -\ 50 \\ \hline \end{array}$

13. $\begin{array}{r} 729 \\ -\ 135 \\ \hline \end{array}$

14. $\begin{array}{r} 225 \\ -\ 133 \\ \hline \end{array}$

15. $\begin{array}{r} 485 \\ -\ 194 \\ \hline \end{array}$

16. $\begin{array}{r} 529 \\ -\ 395 \\ \hline \end{array}$

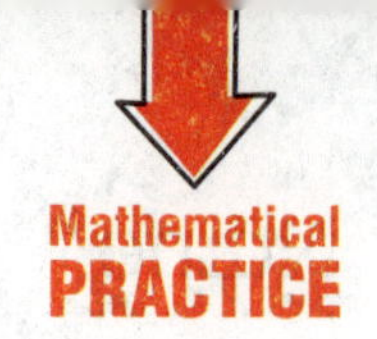

Problem Solving

17. Lucy went to basketball camp. 325 second graders and 234 first graders were there. How many more second graders were at camp?

_______ more second graders

18. 417 boys were at the museum. 286 girls were at the museum. How many more boys were at the museum than girls?

A macaroni artist masterpiece.

_______ more boys

19. 475 parents came to the school's fall party. 294 parents had to leave early. How many parents are still at the fall party?

_______ parents

Write Math Explain how to regroup hundreds.

____

____

____

Name

My Homework

Lesson 5
Regroup Hundreds

Homework Helper

Need help? connectED.mcgraw-hill.com

Find 436 − 245.

Step 1 Subtract ones.

Step 2 Subtract the tens. You cannot subtract 4 from 3. Regroup 1 hundred as 10 tens.

Step 3 Subtract the hundreds.

	hundreds	tens	ones
	3	13	
	~~4~~	~~3~~	6
−	2	4	5
	1	9	1

Practice

Subtract.

1.

	hundreds	tens	ones
	☐	☐	
	3	6	2
−	2	7	1

2.

	hundreds	tens	ones
	☐	☐	
	7	3	2
−	4	4	1

3.

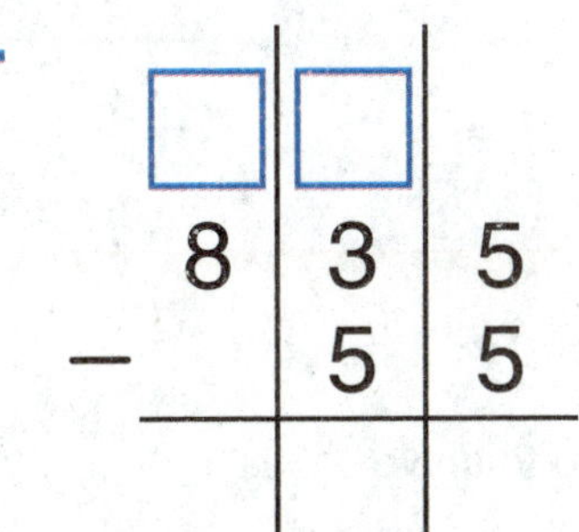

	☐	☐	
	8	3	5
−		5	5

4.

	☐	☐	
	7	3	5
−	6	7	2

5.

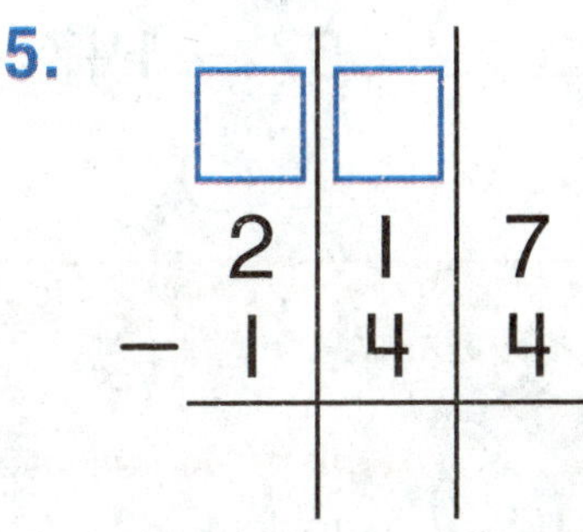

	☐	☐	
	2	1	7
−	1	4	4

Subtract.

6. $\begin{array}{r} 453 \\ -\ 62 \\ \hline \end{array}$	7. $\begin{array}{r} 721 \\ -\ 131 \\ \hline \end{array}$	8. $\begin{array}{r} 745 \\ -\ 552 \\ \hline \end{array}$	9. $\begin{array}{r} 375 \\ -\ 292 \\ \hline \end{array}$
10. $\begin{array}{r} 423 \\ -\ 282 \\ \hline \end{array}$	11. $\begin{array}{r} 434 \\ -\ 243 \\ \hline \end{array}$	12. $\begin{array}{r} 625 \\ -\ 462 \\ \hline \end{array}$	13. $\begin{array}{r} 278 \\ -\ 184 \\ \hline \end{array}$

14. 835 cows are in the field. 251 cows go in the barn. How many cows are still in the field?

______ cows

Test Practice

15. Which subtraction problem does not need regrouping to solve?

363 − 148 ○

734 − 371 ○

357 − 147 ○

367 − 288 ○

Math at Home Have your child explain when you need to regroup to solve a problem.

Name

Subtract Three-Digit Numbers

Lesson 6

ESSENTIAL QUESTION
How can I subtract three-digit numbers?

Explore and Explain

_______ students

Teacher Directions: Use base-ten blocks to solve. 355 students are eating lunch in the cafeteria. 166 students packed their lunch. How many students bought their lunch? Draw the blocks. Write the number.

See and Show

I've got this one!

Find 634 – 159.

Step 1 Subtract the ones. You cannot subtract 9 ones from 4 ones. Regroup 1 ten as 10 ones.

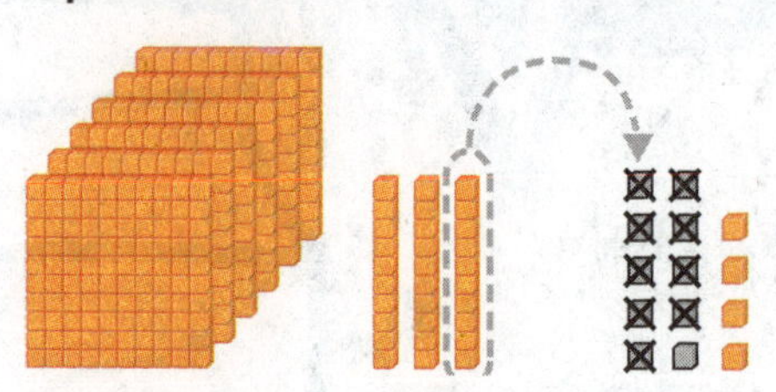

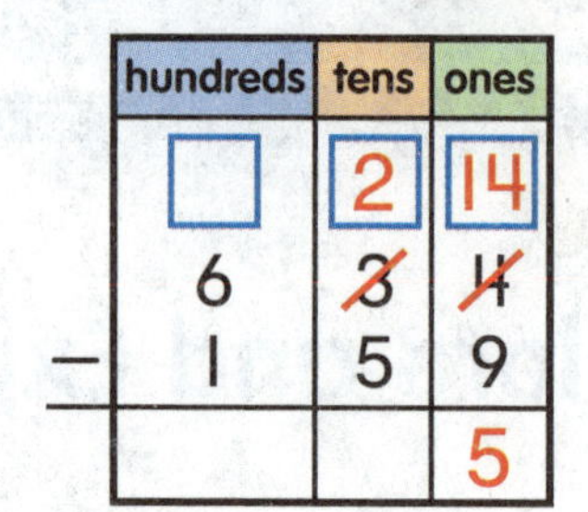

	hundreds	tens	ones
		2	14
	6	~~3~~	~~4~~
–	1	5	9
			5

Step 2 Subtract the tens. You cannot subtract 5 tens from 2 tens. Regroup 1 hundred as 10 tens.

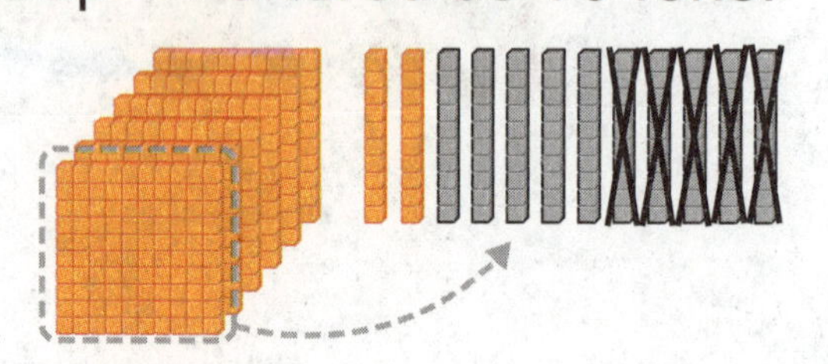

	hundreds	tens	ones
	5	12	14
	~~6~~	~~3~~	~~4~~
–	1	5	9
		7	5

Step 3 Subtract the hundreds. Now you can subtract 1 hundred from 5 hundreds.

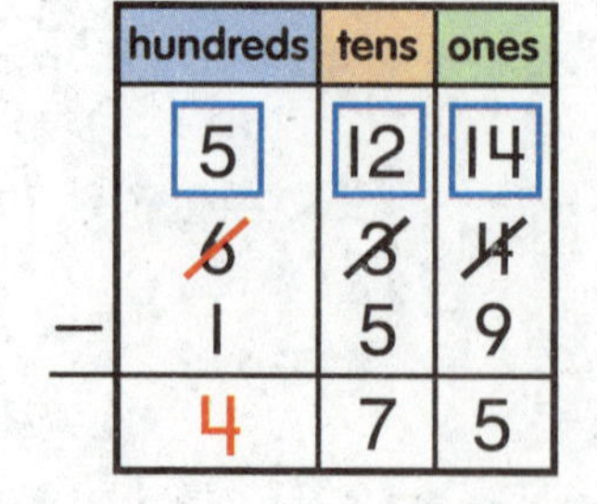

	hundreds	tens	ones
	5	12	14
	~~6~~	~~3~~	~~4~~
–	1	5	9
	4	7	5

634 – 159 = 475

Talk Math Explain what you write in the box above the ones and the box above the tens when you regroup tens and hundreds.

Name

On My Own

Use Work Mat 7 and base-ten blocks. Subtract.

1.
4	3	1
− 3	4	5

2.
7	6	2
−	8	8

3.
9	5	3
− 7	6	4

4. $702 - 211$

5. $884 - 197$

6. $632 - 444$

7. $485 - 296$

8. $357 - 169$

9. $625 - 438$

10. $590 - 184$

11. $718 - 628$

12. $394 - 185$

13. $561 - 273$

14. $934 - 395$

15. $533 - 203$

Problem Solving

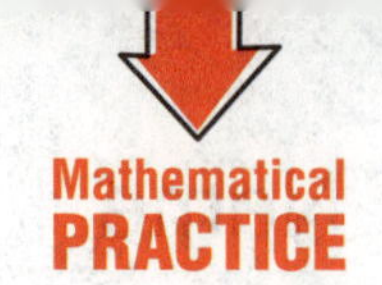

16. The post office had 912 stamps on Friday. By Saturday 189 stamps were left. How many stamps were sold on Friday?

_______ stamps

17. Dakota had 200 plastic bugs. He lost some bugs. He has 155 bugs left. How many bugs did he lose?

_______ bugs

18. There are 350 pieces of popcorn. Ayana and Gabby eat 177 pieces. How many pieces are left?

_______ pieces

HOT Problem 228 students like red. 293 students like blue. 154 students like green. How many more students like red or green than blue? Explain.

____

____

____

Name

My Homework

Lesson 6

Subtract Three-Digit Numbers

Homework Helper

Need help? connectED.mcgraw-hill.com

Sometimes when you subtract, you need to regroup the tens and hundreds.

Find 621 − 475.

Step 1 Subtract the ones. You cannot subtract 5 from 1. Regroup 1 ten as 10 ones.

Step 2 Subtract the tens. You cannot subtract 7 from 1. Regroup.

Step 3 Subtract the hundreds.

	hundreds	tens	ones
	5	11	11
	~~6~~	~~2~~	~~1~~
−	4	7	5
	1	4	6

Practice

Subtract.

1.

	☐	☐	☐
	4	5	3
−	3	5	1

2.

	☐	☐	☐
	6	1	2
−	1	5	9

Subtract.

3.

5	3	7
−	2	6

4.

6	3	4
− 2	7	8

5.

3	6	4
− 1	7	2

6. $\begin{array}{r} 634 \\ -\ \ 26 \\ \hline \end{array}$

7. $\begin{array}{r} 264 \\ -\ 168 \\ \hline \end{array}$

8. $\begin{array}{r} 524 \\ -\ 445 \\ \hline \end{array}$

9. $\begin{array}{r} 347 \\ -\ 168 \\ \hline \end{array}$

10. 344 people are at a game. 198 people leave at half time. How many people are still at the game?

________ people

Test Practice

11. Which problem needs to be regrouped twice?

354 − 134

367 − 263
○

364 − 274

364 − 278
○

Math at Home Have your child explain how regrouping tens and hundreds are the same.

Name

Number and Operations in Base Ten

2.NBT.7

Rewrite Three-Digit Subtraction

Lesson 7

ESSENTIAL QUESTION

How can I subtract three-digit numbers?

Explore and Explain

385 – 266

	hundreds	tens	ones
–			

_____ pages

Teacher Directions: Our teacher is reading a very long book to our class. The book has 385 pages. So far, she has read 266 pages of the book. How many pages does she have left to read? Write the numbers in the place-value chart and subtract.

See and Show

You can rewrite a problem to subtract.
Find 368 − 279.

Step 1 Rewrite.

Step 2 Subtract.

	2	15	18
	~~3~~	~~6~~	~~8~~
−	2	7	9
		8	9

Helpful Hint
Write the greater number at the top. Write the other number below it.

Line up the ones, tens, and hundreds.

Rewrite the problem. Subtract.

1. 336 − 272

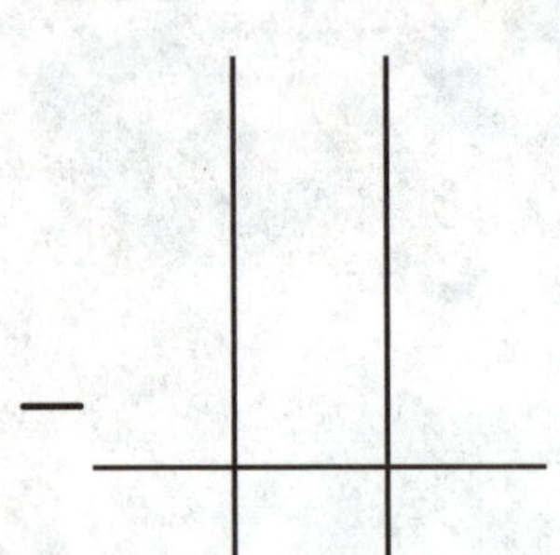

2. 377 − 264

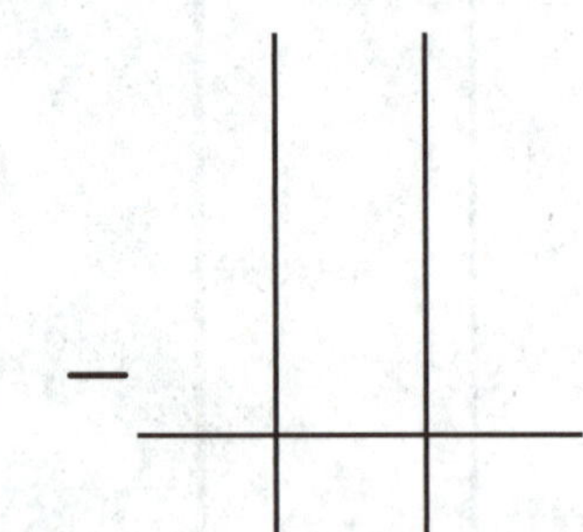

3. 633 − 265

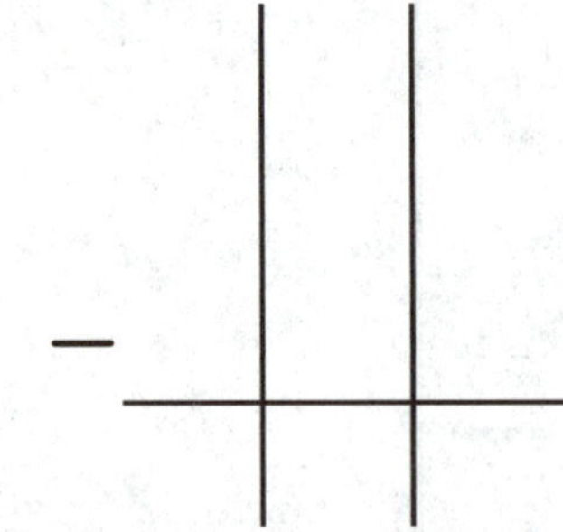

4. 264 − 175

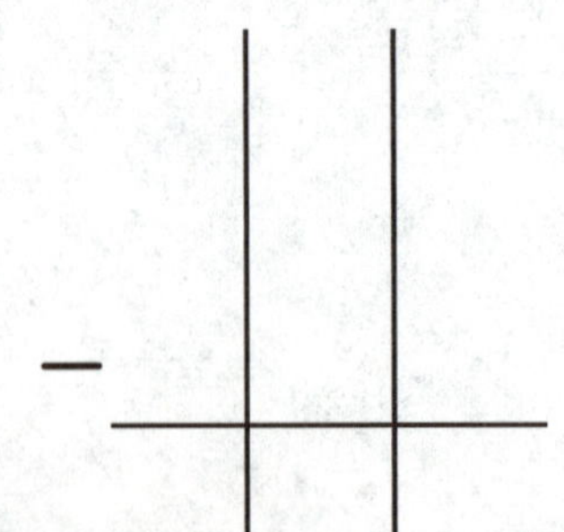

5. 845 − 378

6. 555 − 428

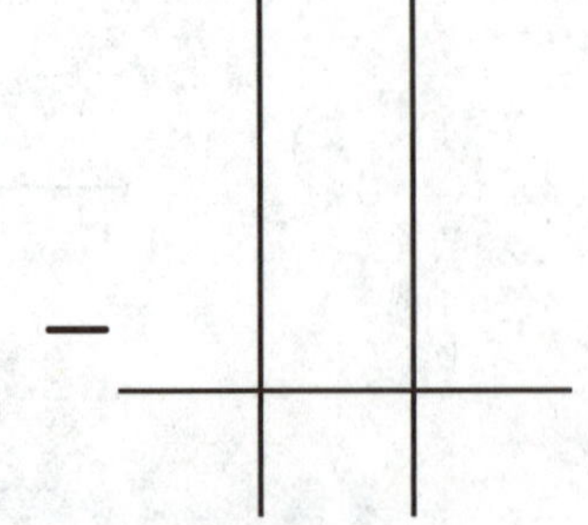

Talk Math How is rewriting three-digit subtraction different than when you rewrite two-digit subtraction?

Name ....

On My Own

Rewrite the problem. Subtract.

7. 363 − 278

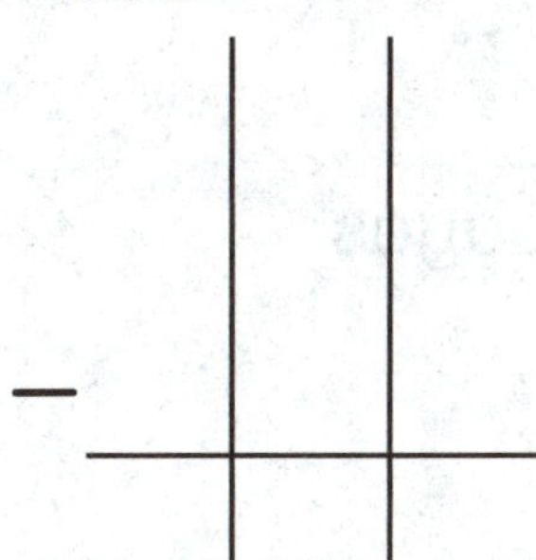

8. 285 − 185

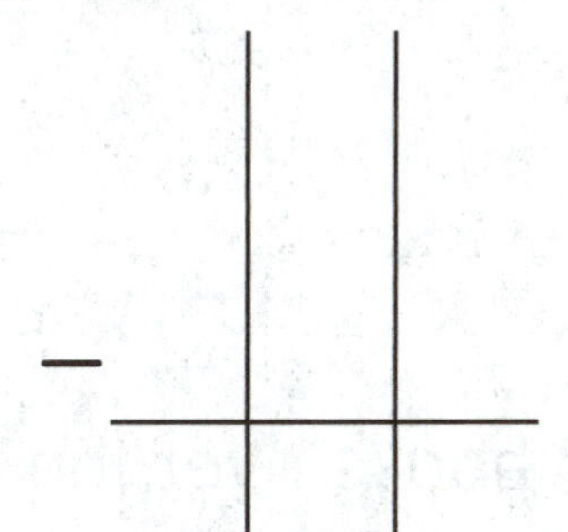

9. 634 − 175

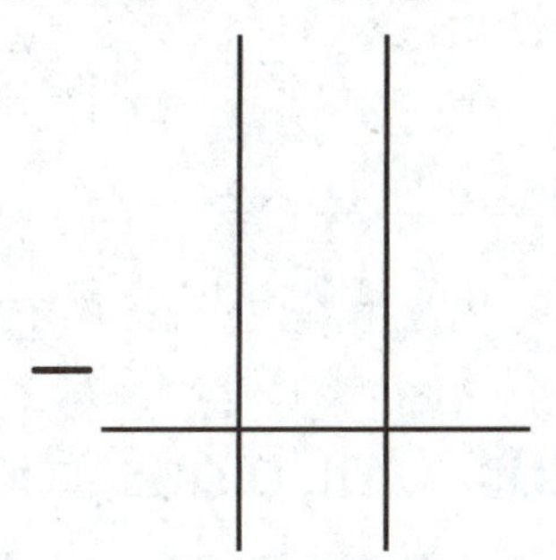

10. 375 − 142

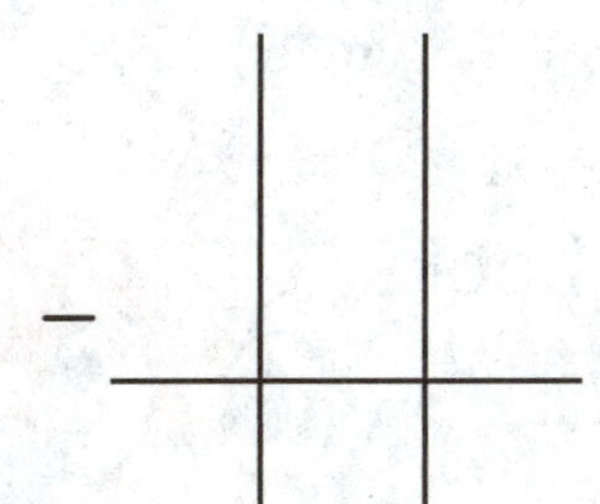

11. 825 − 195

12. 647 − 373

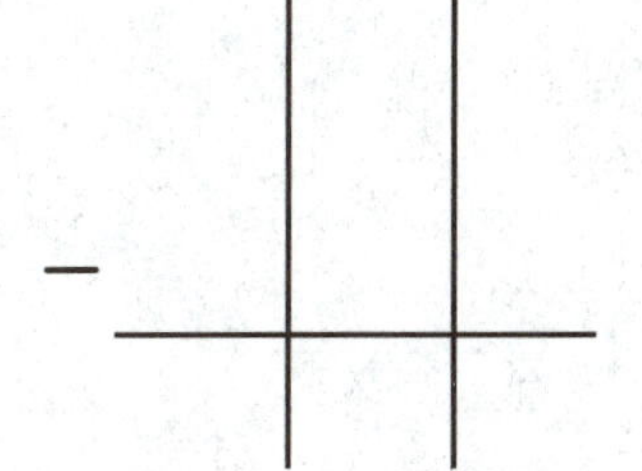

13. 695 − 295

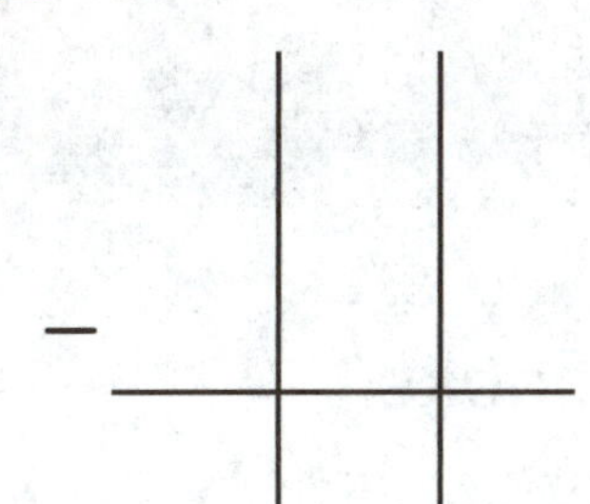

14. 853 − 259

−

15. 496 − 349

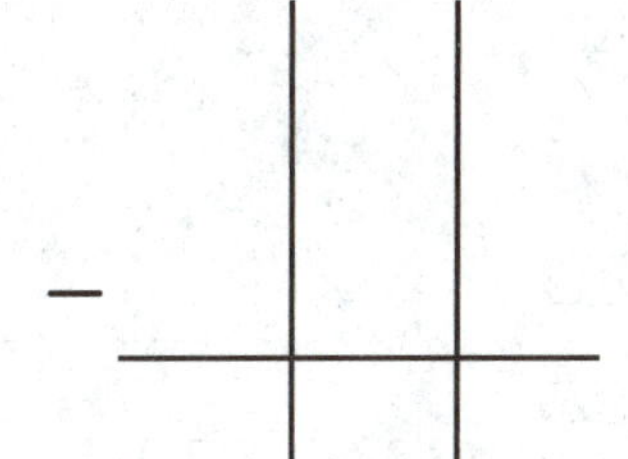

16. 495 − 267

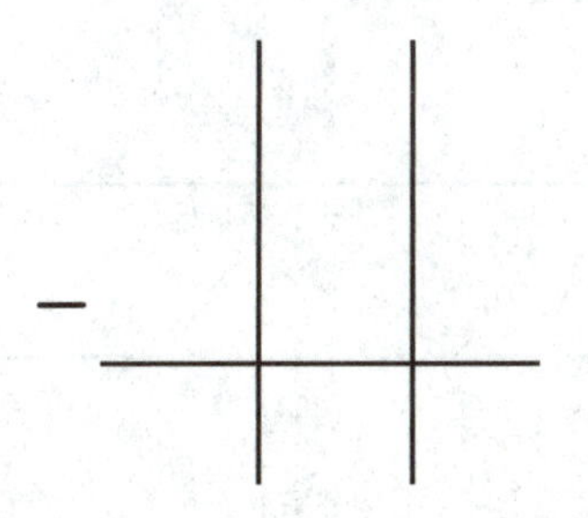

17. 845 − 264

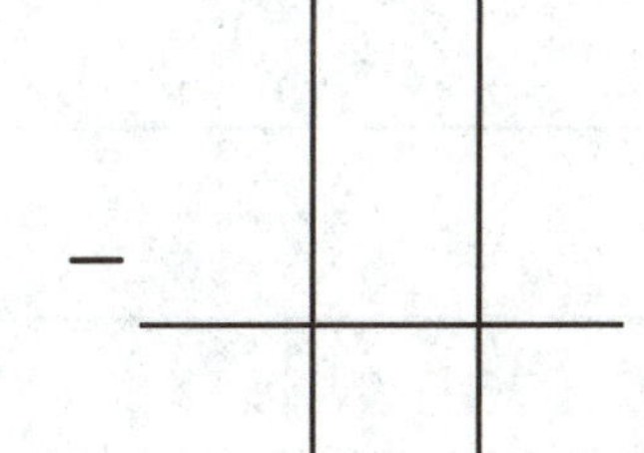

18. 764 − 375

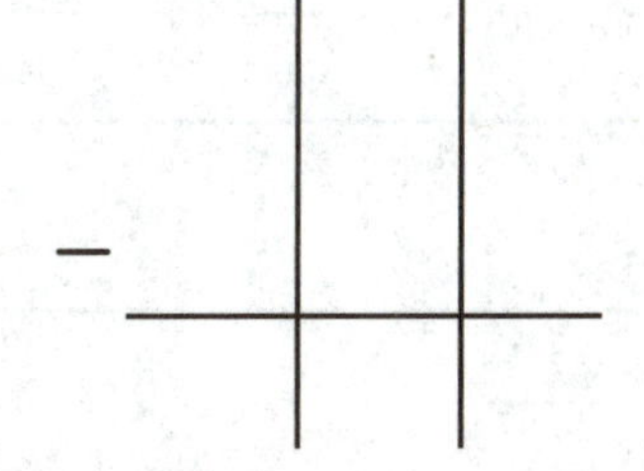

Problem Solving

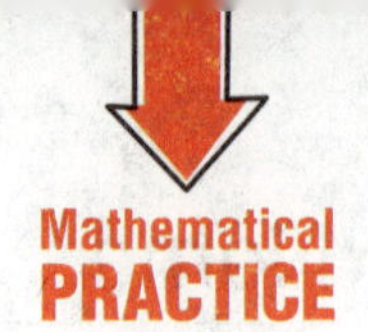

19. Lucas is reading a book that is 239 pages long. He has read 159 pages. How many pages does he have left to read?

______ pages

20. Our class read 753 books over the summer. The girls read 394 books. How many books did the boys read?

I do a lot of summer reading!

______ books

21. Our school library has 125 books about pets. I have read 96 of the books. How many books do I have left to read?

______ books

Write Math Explain how you rewrite three-digit subtraction problems to solve.

____

____

____

Name ....

My Homework

Lesson 7
Rewrite Three-Digit Subtraction

Homework Helper

Need help? connectED.mcgraw-hill.com

Find 356 − 298.

Step 1 Rewrite. Place the greater number on top.

Step 2 Subtract. Regroup if necessary.

	2	14	16
	~~3~~	~~5~~	~~6~~
−	2	9	8
		5	8

Practice

Rewrite the problem. Subtract.

1. 724 − 235

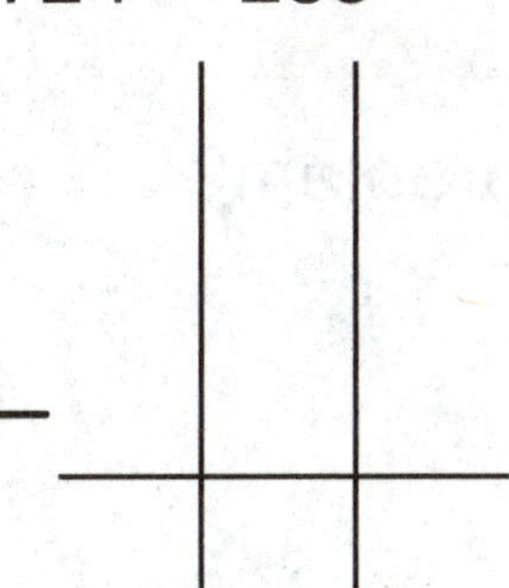

2. 616 − 337

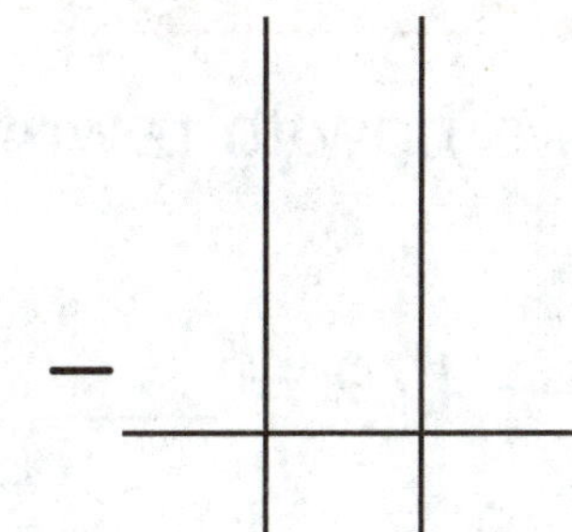

3. 374 − 286

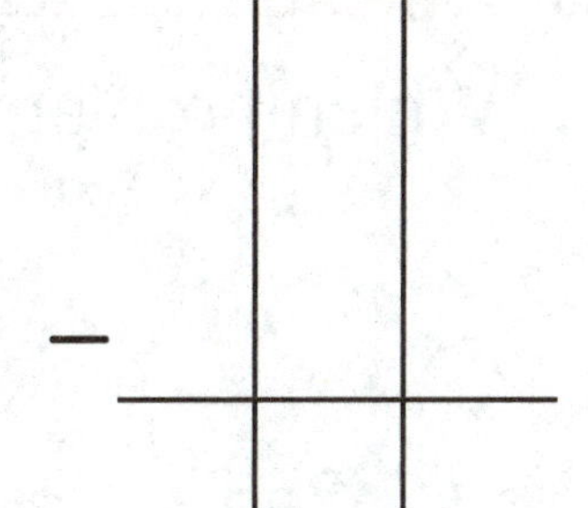

4. 875 − 596

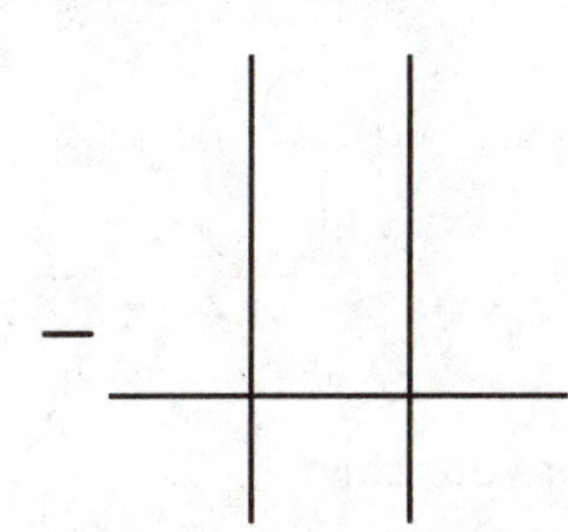

5. 945 − 387

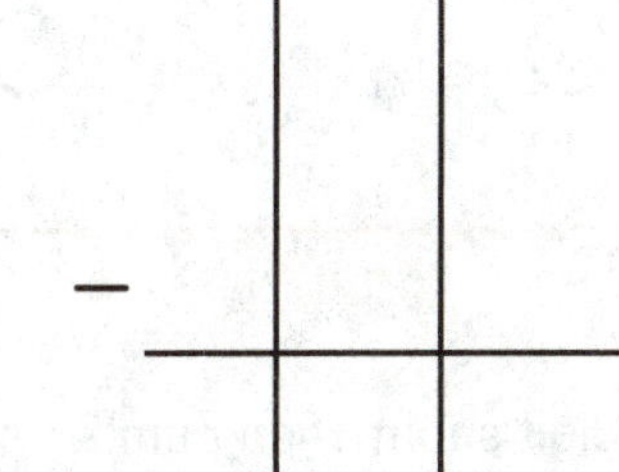

6. 435 − 294

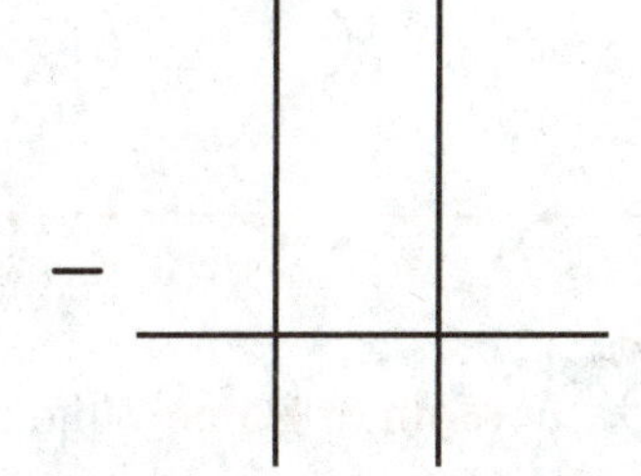

Rewrite the problem. Subtract.

7. 162 – 89

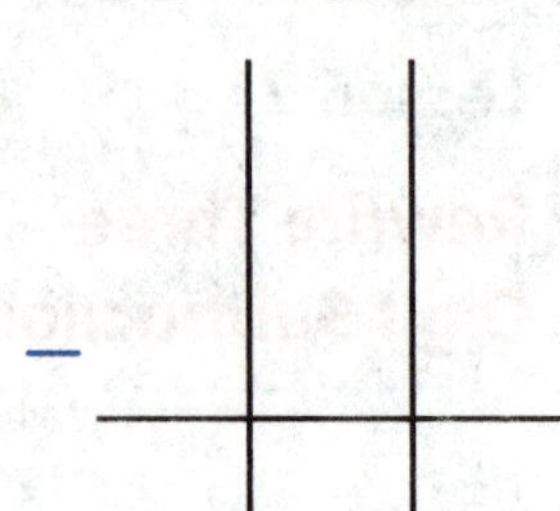

8. 619 – 254

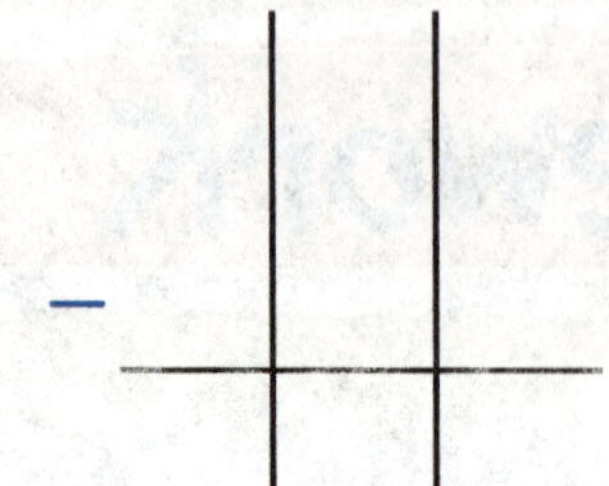

9. 195 – 99

10. 835 people are on a boat. 295 people get off of the boat. How many people are still on the boat?

_______ people

Test Practice

11. Which problem shows how to rewrite the problem?

368 – 179 = _______

368 – 179 200	179 – 368 411	368 – 179 189	368 + 179 547
○	○	○	○

Math at Home Write a three-digit subtraction number sentence for your child. Have him or her rewrite the number sentence vertically and then subtract.

Name ....

Problem Solving

STRATEGY: Write a Number Sentence

Lesson 8

ESSENTIAL QUESTION
How can I subtract three-digit numbers?

Fido always begs for treats. He has a bag of 255 treats. By the end of the month, he had eaten 82 treats. How many treats are left for Fido to eat?

Watch

1 Understand
Underline what you know.
Circle what you need to find.

2 Plan
How will I solve the problem?

3 Solve
Write a number sentence.

255 − 82 = 173 treats are left

4 Check
Is my answer reasonable? Explain.

Practice the Strategy

385 students entered the science fair. 193 of the students were girls. How many students were boys?

1 Understand Underline what you know.
Circle what you need to find.

2 Plan How will I solve the problem?

3 Solve I will...

_____ − _____ = _____

4 Check Is my answer reasonable? Explain.

Name ______________________

CCSS

Apply the Strategy

1. It took 227 days to build railroad tracks over a mountain. It took 132 days to build tracks over flat ground. How many more days did it take to build over the mountain?

_____ ○ _____ ○ _____ days

2. The Fuller family is driving 475 miles. They have already gone 218 miles. How many miles are left to go?

_____ ○ _____ ○ _____ miles

3. Flora's class is trying to collect 850 cans. She has turned in 370 cans. How many cans are needed to reach the class goal?

_____ ○ _____ ○ _____ cans

Review the Strategies

Choose a strategy

- Make a model.
- Guess, check, and revise.
- Write a number sentence.

4. Anna has been keeping track of the weather for 289 days. 196 of the days were sunny. How many days were not sunny?

______ days

5. There are 836 flowers in the field. The children pick 398 of the flowers. How many flowers are left in the field?

______ flowers

6. 423 fish are in the tank. 184 fish are removed from the tank. How many fish are in the tank now?

______ fish

Name

My Homework

Lesson 8
Problem Solving: Write a Number Sentence

638 people started the race. 459 people finished the race. How many people did not finish the race?

1 Understand

Underline what you know.
Circle what you need to find.

2 Plan

How will I solve the problem?

3 Solve

Write a number sentence.

638	–	459	=	179
people started		people finished		people did not finish

179 people did not finish the race.

4 Check

Is my answer reasonable?

Underline what you know. Circle what you need to find.

1. A farmer picks 389 ears of corn. He sells 183 ears of corn. How many ears of corn does the farmer have left?

______ ◯ ______ ◯ ______ ears of corn

2. 135 people are at the park. 46 people are playing baseball. How many people are not playing baseball?

I'll take 2 hot dogs with mustard!

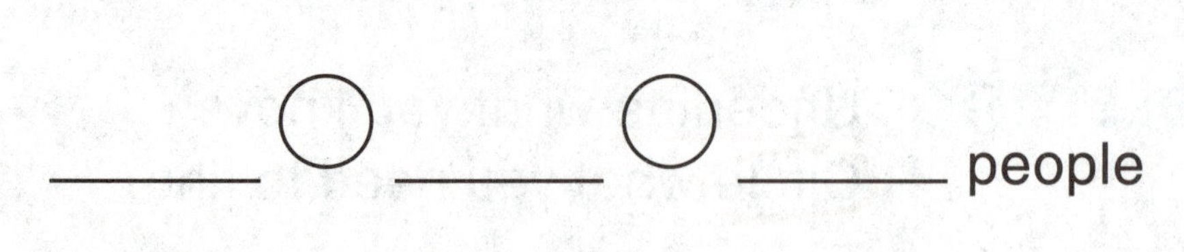

3. There are 276 animals at the zoo. 185 of the animals are male. How many animals are female?

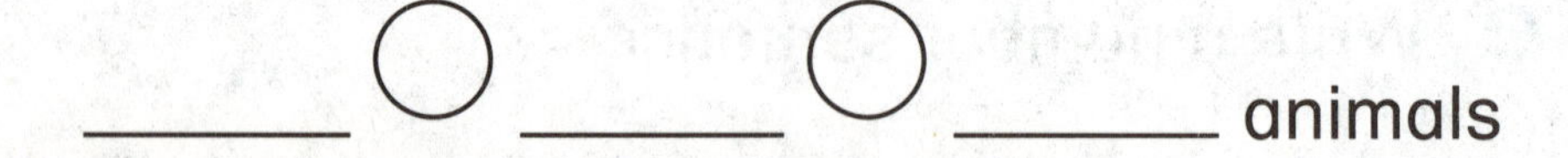

Test Practice

4. 152 tomatoes grew in my garden last year. This year only 98 tomatoes grew. How many more tomatoes grew last year?

252	250	54	52
◯	◯	◯	◯

Math at Home Have your child find 477 – 293 by writing a number sentence.

Name

Subtract Across Zeros

Lesson 9

ESSENTIAL QUESTION
How can I subtract three-digit numbers?

Explore and Explain

Yippee, I'm a mammal!

Me too!

_____ animals

Teacher Directions: Model using base-ten blocks. We have learned about 200 different kinds of animals. 126 of those animals were mammals. How many of the animals were not mammals? Draw the blocks. Write the number.

See and Show

Mathematical PRACTICE

You can subtract across zeros.

Find 400 − 234.

Step 1 Subtract the ones. You cannot subtract 4 ones from 0 ones. There are no tens to subtract from. Look at the hundreds. Regroup 1 hundred as 10 tens. Then regroup 1 ten as ten ones.

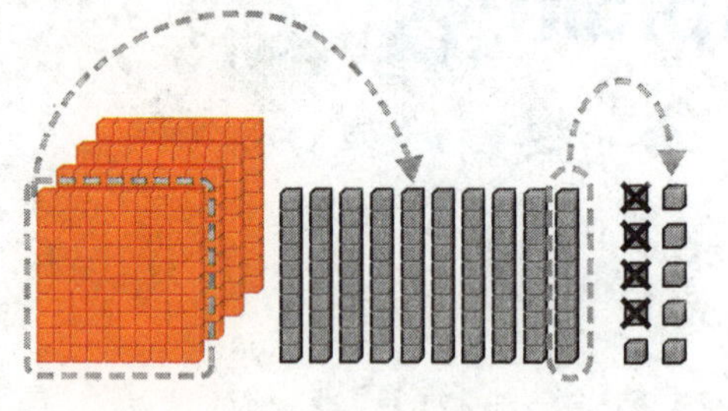

	hundreds	tens	ones
			10
	~~4~~	~~0~~	~~0~~
−	2	3	4
			6

Step 2 Subtract the tens. There are 9 tens left. Subtract 3 tens from 9 tens.

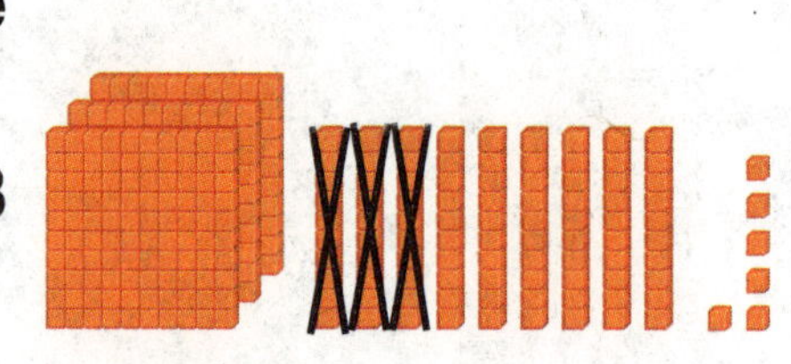

	hundreds	tens	ones
		9	10
	~~4~~	~~0~~	~~0~~
−	2	3	4
		6	6

Step 3 Subtract the hundreds. There are 3 hundreds left. Subtract 2 hundreds from 3 hundreds.

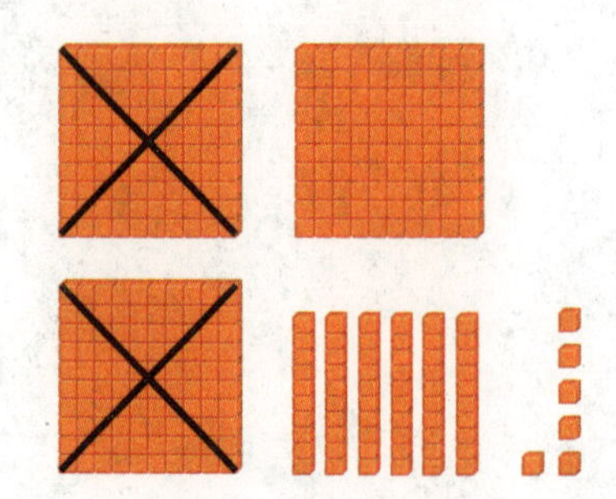

	hundreds	tens	ones
	3	9	10
	~~4~~	~~0~~	~~0~~
−	2	3	4
	1	6	6

400 − 234 = 166

Talk Math How is subtracting from 400 different than subtracting from 435?

Name

On My Own

Use Work Mat 7 and base-ten blocks. Subtract.

1.

hundreds	tens	ones
☐	☐	☐
8	0	0
− 5	3	2

2.

hundreds	tens	ones
☐	☐	☐
7	0	0
− 6	1	4

3. $100 - 76$

4. $900 - 287$

5. $400 - 167$

6. $700 - 444$

7. $300 - 16$

8. $800 - 477$

9. $900 - 876$

10. $500 - 54$

11. $800 - 691$

12. $200 - 75$

13. $500 - 321$

14. $600 - 312$

Problem Solving

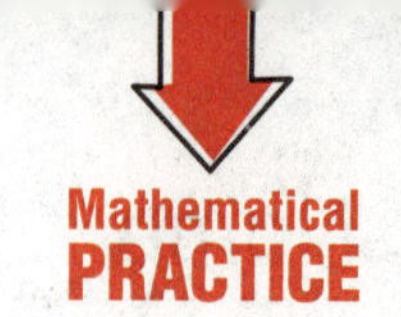

15. 400 bees are in the hive. 145 bees leave the hive. How many bees are still in the hive?

_____ bees

16. 300 girls took gymnastics last year. This year 193 girls take gymnastics. How many more girls took gymnastics last year than this year?

_____ girls

HOT Problem 500 people signed up for soccer. 123 people stopped playing. Then 154 more people stopped playing. How many people still play soccer? Explain.

____

____

____

Name

My Homework

Lesson 9
Subtract Across Zeros

Homework Helper

Need help? connectED.mcgraw-hill.com

You can subtract across zeros.
Find 600 − 336.

Step 1 Subtract the ones. Regroup 1 hundred into 10 tens. Then regroup 1 ten as 10 ones.

Step 2 Subtract the tens.

Step 3 Subtract the hundreds.

	hundreds	tens	ones
	5	9	10
	~~6~~	~~0~~	~~0~~
−	3	3	6
	2	6	4

Practice

Subtract.

1. 300 − 251

2. 600 − 139

3. 700 − 386

4. 200 − 126

5. 500 − 385

6. 800 − 272

Subtract.

7. $100 - 89$

8. $600 - 564$

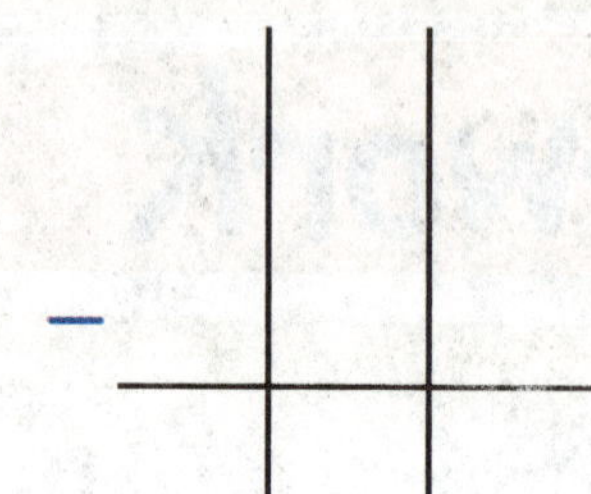

9. $500 - 268$

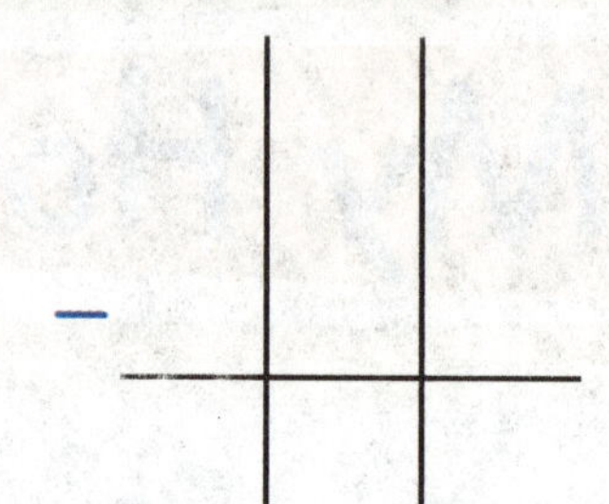

10.
$$\begin{array}{r} 900 \\ -\ 432 \\ \hline \end{array}$$

11.
$$\begin{array}{r} 700 \\ -\ 364 \\ \hline \end{array}$$

12.
$$\begin{array}{r} 200 \\ -\ 147 \\ \hline \end{array}$$

13. 400 pineapples are planted in a row. 293 pineapples get picked. How many pineapples are still in the row?

_____ pineapples

I spiked my hair!

Test Practice

14. Find $600 - 289$.

889	311	301	489
○	○	○	○

Math at Home Have your child explain how to subtract 392 from 800.

Name ....

My Review

Chapter 7
Subtract Three-Digit Numbers

Vocabulary Check 

place value **regroup** **subtract**

Write the correct word in the blank.

1. To identify the value of the digit 9 in the numbers 976, 93, and 9, use what you know about ______________.

2. You can ______________ to find the difference.

3. To take a number apart and write it a new way is to ______________.

Concept Check 

Subtract.

4. 600 − 400 = ______

5. 700 − 600 = ______

Subtract.

6. $\begin{array}{r} 800 \\ -\ 400 \\ \hline \end{array}$

7. $\begin{array}{r} 900 \\ -\ 600 \\ \hline \end{array}$

8. $\begin{array}{r} 773 \\ -\ 100 \\ \hline \end{array}$

9. $\begin{array}{r} 261 \\ -\ 10 \\ \hline \end{array}$

10. $\begin{array}{r} 938 \\ -\ 329 \\ \hline \end{array}$

11. $\begin{array}{r} 885 \\ -\ 16 \\ \hline \end{array}$

12. $\begin{array}{r} 357 \\ -\ 189 \\ \hline \end{array}$

13. $\begin{array}{r} 987 \\ -\ 598 \\ \hline \end{array}$

14. $\begin{array}{r} 201 \\ -\ 124 \\ \hline \end{array}$

Rewrite the problem. Subtract.

15. 385 − 166

−

16. 247 − 189

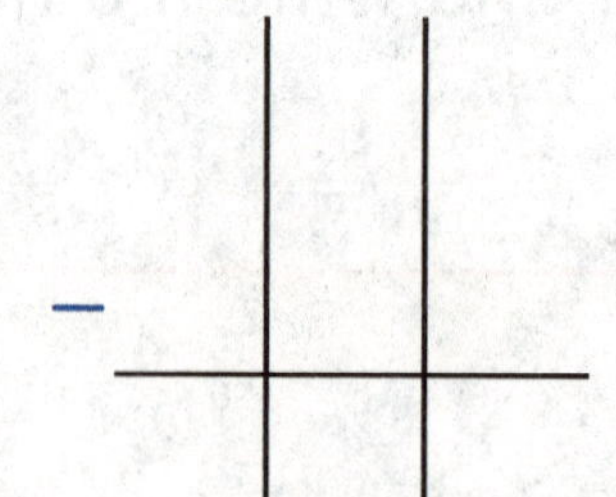

17. 925 − 638

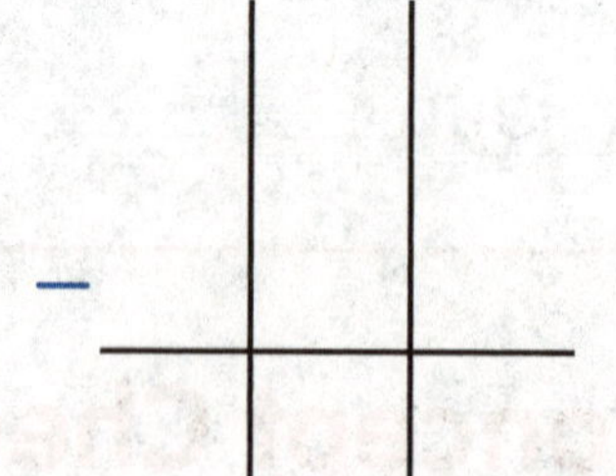

Subtract.

18. $\begin{array}{r} 400 \\ -\ 254 \\ \hline \end{array}$

19. $\begin{array}{r} 700 \\ -\ 443 \\ \hline \end{array}$

20. $\begin{array}{r} 300 \\ -\ 165 \\ \hline \end{array}$

Name ....

Problem Solving

I ♥ buttons!

21. There are 620 buttons. 200 of them are square. The rest are round. How many buttons are round?

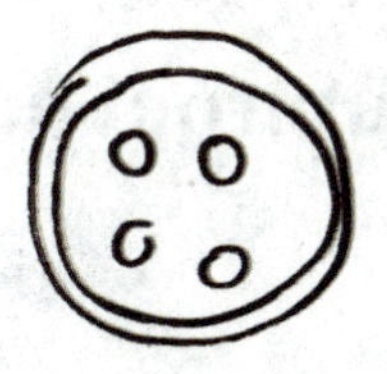

_______ buttons

22. Clara had 359 pennies. She lost some. Now she has 266 pennies. How many pennies did Clara lose?

_______ pennies

Test Practice

23. There are 334 cars in the parking lot on Sunday. On Monday, there are 182 cars. How many more cars were there on Sunday?

334	182	152	150
○	○	○	○

Reflect

Chapter 7

Answering the Essential Question

Show the ways to subtract three-digit numbers.

Subtract hundreds.	Regroup tens to subtract.
800 − 500 = ______	$\begin{array}{r} 835 \\ -\ 726 \\ \hline \end{array}$
Regroup tens and hundreds to subtract.	Rewrite to subtract.
$\begin{array}{r} 935 \\ -\ 397 \\ \hline \end{array}$	381 − 298

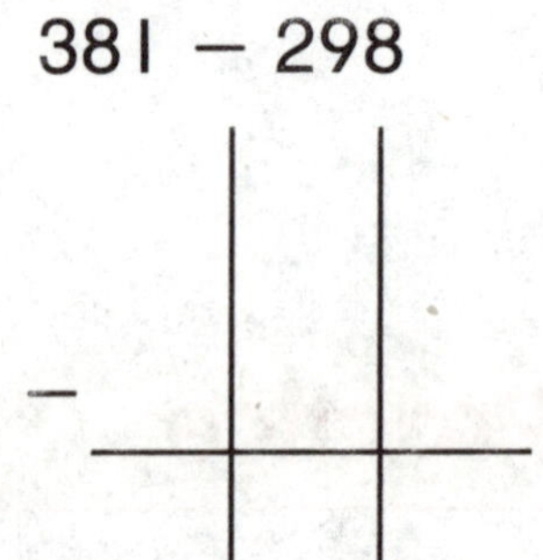

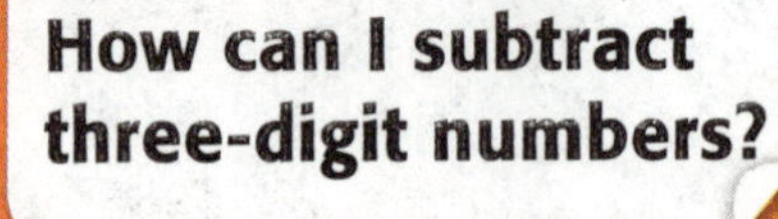

ESSENTIAL QUESTION

How can I subtract three-digit numbers?

You are part of this equation!

You + School = Cool

Glossary/Glosario

Go online for the eGlossary.

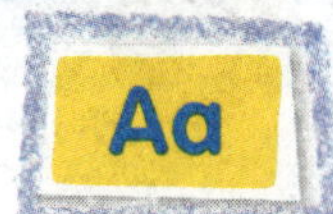

English	Spanish/Español
A.M. The hours from midnight until noon.	**a.m.** Las horas que van desde la medianoche hasta el mediodía.
add (addition) Join together sets to find the total or sum. The opposite of *subtract*.	**sumar (adición)** Unir conjuntos para hallar el total o la suma. Lo opuesto de restar.
$2 + 5 = 7$ 	$2 + 5 = 7$ 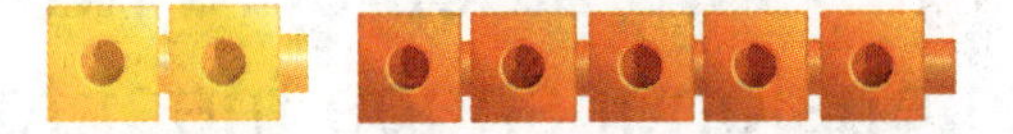
addend Any numbers or quantities being added together.	**sumando** Numeros o cantidades que se suman.
In $2 + 3 = 5$, 2 is an addend and 3 is an addend.	En $2 + 3 = 5$, 2 es un sumando y 3 es un sumando.
$2 + 3 = 5$ (arrows point to 2 and 3)	$2 + 3 = 5$ (arrows point to 2 and 3)

Aa

after Follow in place or time.

5 6 7 8

6 is just *after* 5

después Que sigue en lugar o en tiempo.

5 6 7 8

6 está justo *después* del 5

analog clock A clock that has an hour hand and a minute hand.

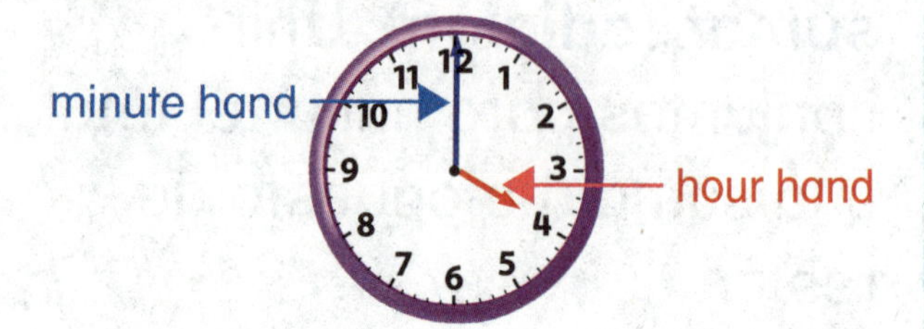

reloj analógico Reloj que tiene una manecilla horaria y un minutero.

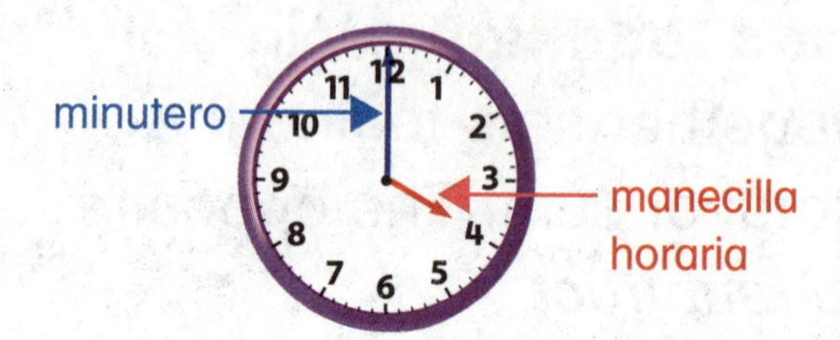

angle Two sides on a two-dimensional shape meet to form an angle.

ángulo Dos lados de una figura bidimensional se encuentran para formar un ángulo.

array Objects displayed in rows and columns.

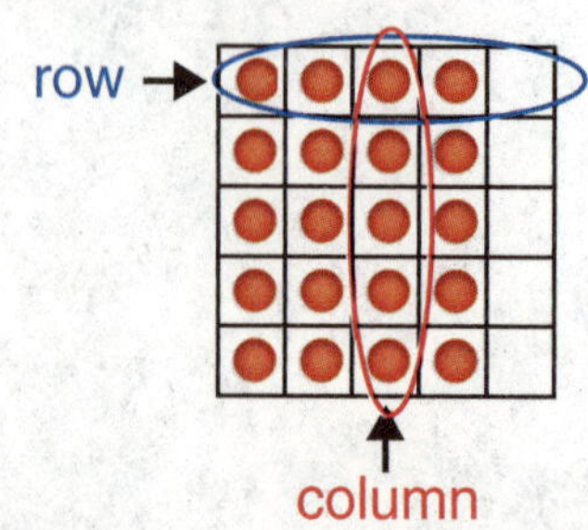

arreglo Objetos organizados en filas y columnas.

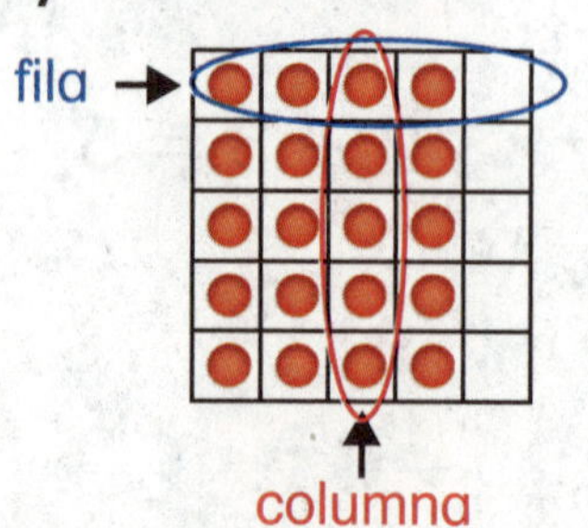

bar graph A graph that uses bars to show data.

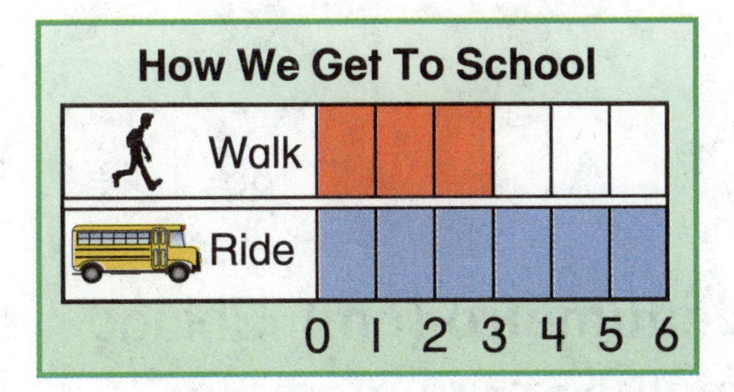

gráfica de barras Gráfica que usa barras para ilustrar datos.

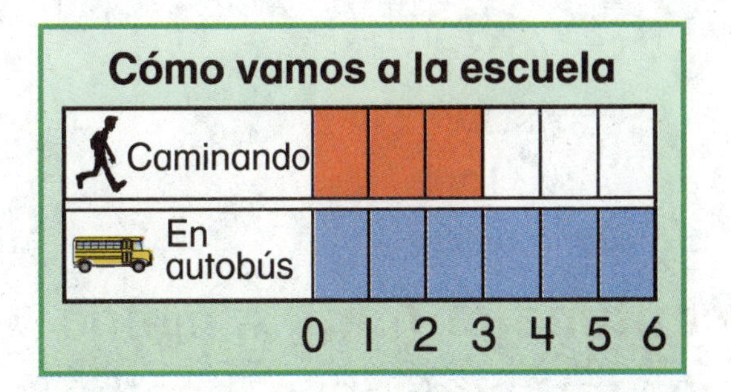

before

5 6 7 8

6 is just *before* 7

antes

5 6 7 8

6 está justo *antes* del 7

between

47 48 49 50

49 is *between* 48 and 50

entre

47 48 49 50

49 está *entre* 48 y 50

cent

1¢ 1 cent

centavo

1¢ 1 centavo

cent sign (¢) The sign used to show cents.

1¢

5¢

signo de centavo (¢) El signo que se usa para mostrar centavos.

1¢

5¢

centimeter (cm) A metric unit for measuring length.

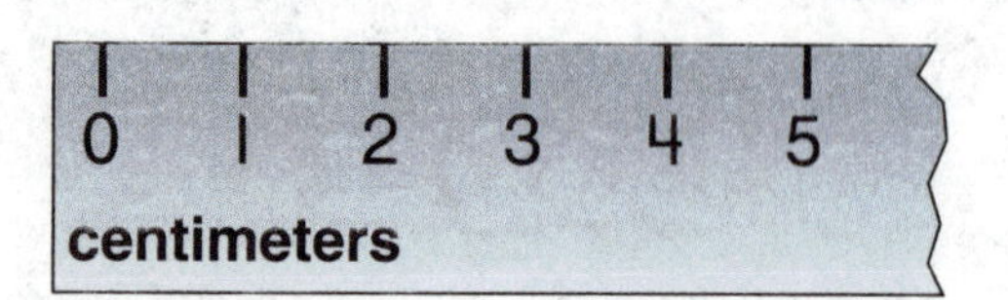

centímetro (cm) Unidad métrica para medir la longitud.

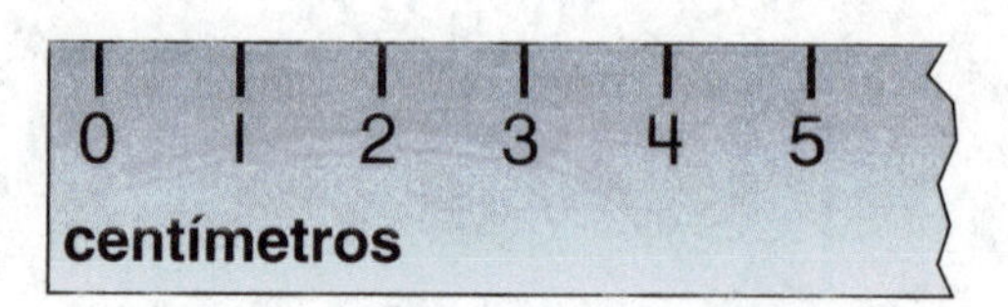

circle A closed, round two-dimensional shape.

círculo Bidimensional redonda y cerrada.

compare Look at objects, shapes, or numbers and see how they are alike or different.

comparar Observar objetos, formas o números para saber en qué se parecen y en qué se diferencian.

cone A three-dimensional shape that narrows to a point from a circular face.

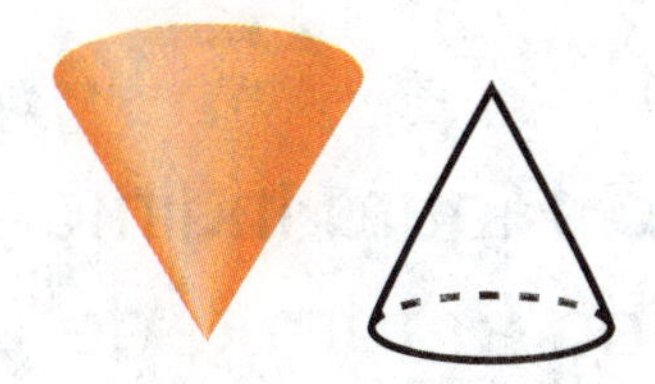

cono Figura tridimensional que se estrecha hasta un punto desde una base circular.

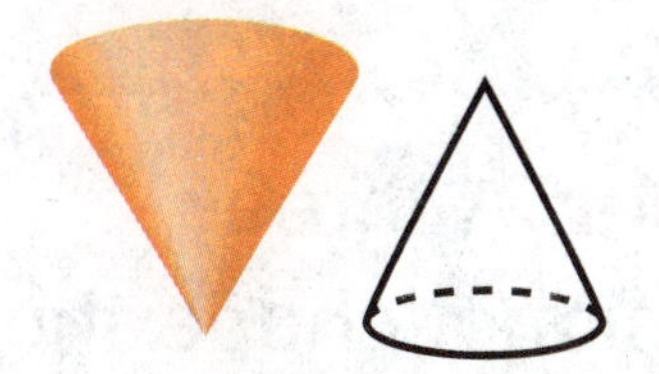

count back On a number line, start at the greater number (5) and count back (3).

$5 - 3 = 2$

2 3 4 **5** 6

contar hacia atrás En una recta numérica, comienza en un número mayor (5) y cuenta (3) hacia atrás.

$5 - 3 = 2$

2 3 4 **5** 6

count on On a number line, start at the first addend (4) and count on (2).

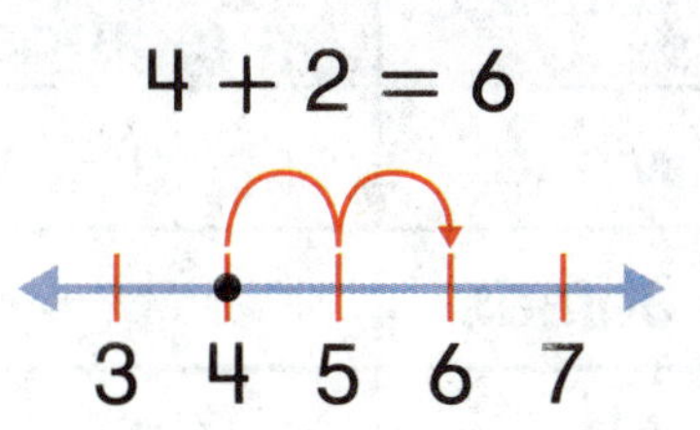

seguir contando En una recta numérica, comienza en el primer sumando (4) y cuenta (2) hacia delante.

$4 + 2 = 6$

3 4 5 6 7

cube A three-dimensional shape with 6 square faces.

cubo Figura tridimensional con 6 caras cuadradas.

cylinder A three-dimensional shape that is shaped like a can.

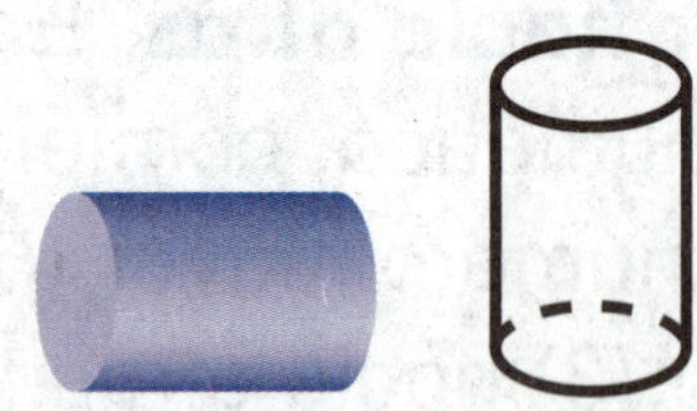

cilindro Figura tridimensional que tiena la forma de una lata.

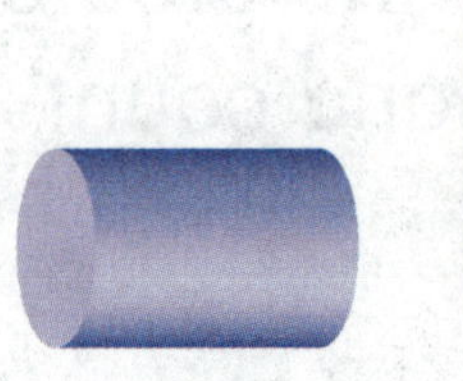
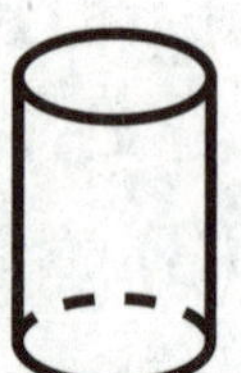

Dd

data Numbers or symbols, sometimes collected from a survey or experiment, that show information. *Data* is plural.

Name	Number of Pets
Mary	3
James	1
Alonzo	4

datos Números o símbolos que se recopilan mediante una encuesta o experimento para mostrar información.

Nombre	Número de mascotas
María	3
James	1
Alonzo	4

day 1 day = 24 hours
Examples: Sunday, Monday, Tuesday, Wednesday, Thursday, Friday, Saturday

día 1 día = 24 horas
Ejemplos: domingo, lunes, martes, miércoles, jueves, viernes y sábado

difference The answer to a subtraction problem.

$$3 - 1 = 2$$

The difference is 2.

diferencia Resultado de un problema de resta.

$$3 - 1 = 2$$

La diferencia es 2.

digit A symbol used to write numbers. The ten digits are

0, 1, 2, 3, 4, 5, 6, 7, 8, 9.

dígito Símbolo que se utiliza para escribir números. Los diez dígitos son
0, 1, 2, 3, 4, 5, 6, 7, 8, 9.

digital clock A clock that uses only numbers to show time.

reloj digital Reloj que marca la hora solo con números.

dime dime = 10¢ or 10 cents

head

tail

moneda de 10¢ moneda de diez centavos = 10¢ o 10 centavos

cara

cruz

dollar one dollar = 100¢ or 100 cents. It can also be written as $1.00 or $1.

front

back

dólar un dólar = 100¢ o 100 centavos. También se puede escribir $1.00.

frente

revés

dollar sign ($) The sign used to show dollars.

one dollar = $1 or $1.00

signo de dólar ($) Símbolo que se usa para mostrar dólares.

un dólar = $1 o $1.00

doubles (and near doubles) Two addends that are the same number.

$6 + 6 = 12$ ← doubles

$6 + 7 = 13$ ← near doubles

dobles (y casi dobles) Dos sumandos que son el mismo número.

$6 + 6 = 12$ ← dobles

$6 + 7 = 13$ ← casi dobles

Ee

edge The line segment where two *faces* of a three-dimensional shape meet.

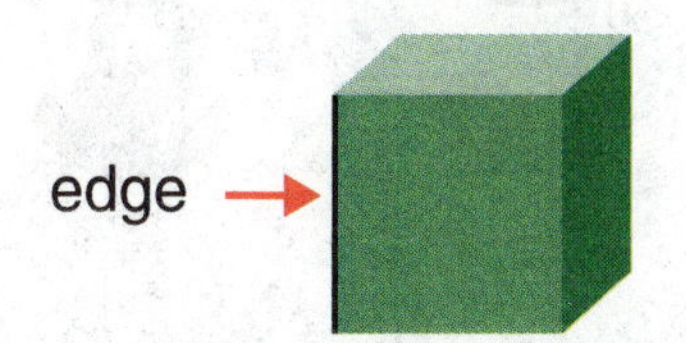

arista Segmento de recta donde se encuentran dos caras de una figura tridimensional.

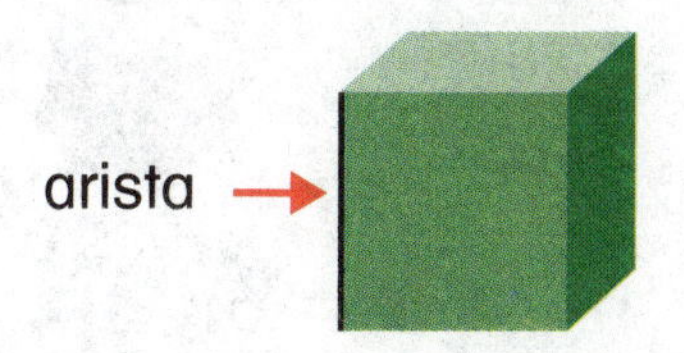

equal groups Each group has the same number of objects.

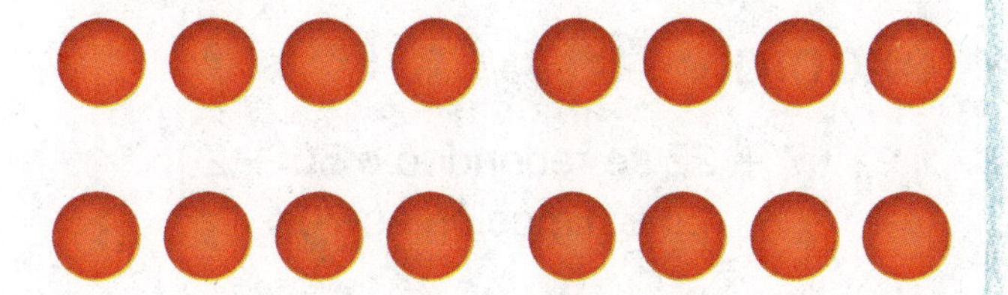

There are four equal groups of counters.

grupos iguales Cada grupo tiene el mismo número de objetos.

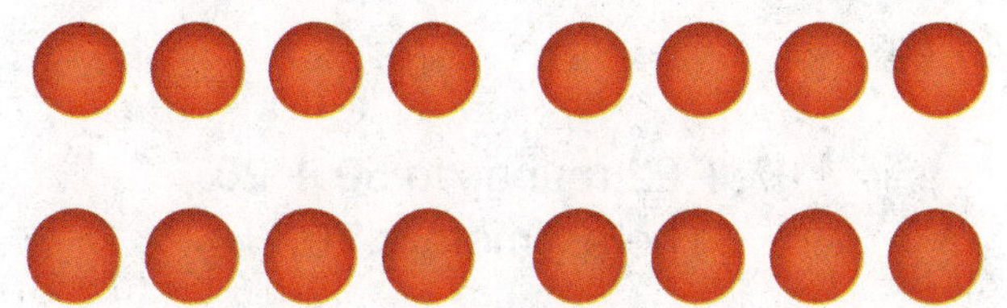

Hay cuatro grupos iguales de fichas.

equal parts Each part is the same size.

This sandwich is cut into 2 equal parts.

partes iguales Cada parte es del mismo tamaño.

Este sándwich está cortado en 2 partes iguales.

Ee

equal to =

6 = 6

6 is *equal to* or the same as 6.

igual a =

6 = 6

6 es *igual* o lo mismo que 6.

estimate Find a number close to an exact amount.

47 + 22 rounds to 50 + 20.
The estimate is 70.

estimar Hallar un número cercano a la cantidad exacta.

47 + 22 se redondea a 50 + 20.
La estimación es 70.

even number Numbers that end with 0, 2, 4, 6, 8.

número par Los números que terminan en 0, 2, 4, 6, 8.

expanded form
The representation of a number as a sum that shows the value of each digit. Sometimes called *expanded notation*.

536 is written as 500 + 30 + 6.

forma desarrollada
La representación de un número como suma que muestra el valor de cada dígito. También se llama *notación desarrollada*.

536 se escribe como 500 + 30 + 6.

face The flat part of a three-dimensional shape.

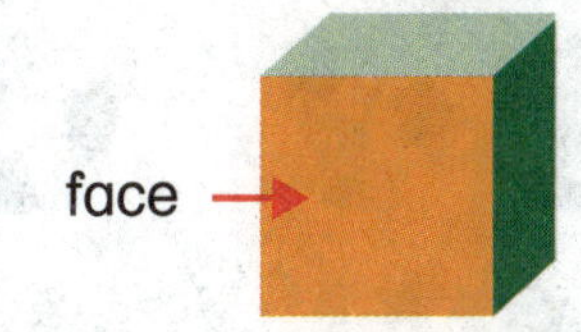

A square is a face of a cube.

cara La parte plana de una figura tridimensional.

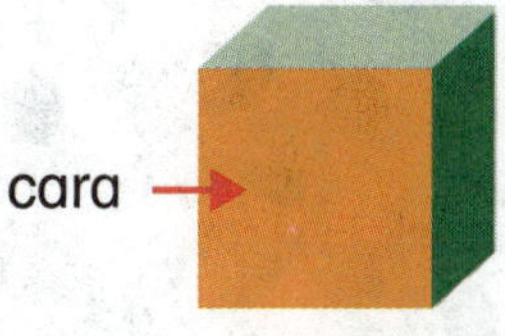

El cuadrado es la cara de un cubo.

fact family Addition and subtraction sentences that use the same numbers.

$6 + 7 = 13$ $\quad 13 - 7 = 6$

$7 + 6 = 13$ $\quad 13 - 6 = 7$

familia de operaciones Enunciados de suma y resta los cuales tienen los mismos números.

$6 + 7 = 13$ $\quad 13 - 7 = 6$

$7 + 6 = 13$ $\quad 13 - 6 = 7$

foot (ft) A customary unit for measuring length. Plural is feet.

1 foot = 12 inches

pie Unidad usual para medir longitud.

1 pie = 12 pulgadas

fourths Four equal parts of a whole. Each part is a fourth, or a quarter of the whole.

cuartos Cuatro partes iguales de un todo. Cada parte es un cuarto, o la cuarta parte del todo.

Gg

greater than >

7 > 2

7 is greater than 2.

mayor que >

7 > 2

7 es mayor que 2.

group A set of objects.

1 group of 4

grupo Conjunto o grupo de objetos.

1 grupo de 4

half hour (or half past) A unit to measure time. Sometimes called *half past* or *half past the hour.*

a half hour = 30 minutes

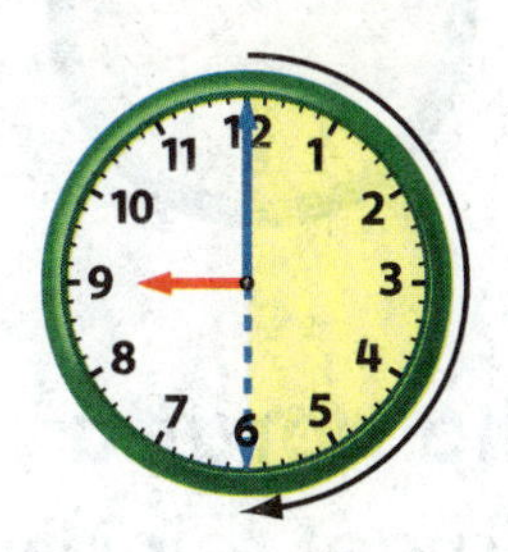

media hora (o y media) Unidad para medir tiempo. A veces se dice *hora y media.*

media hora = 30 minutos

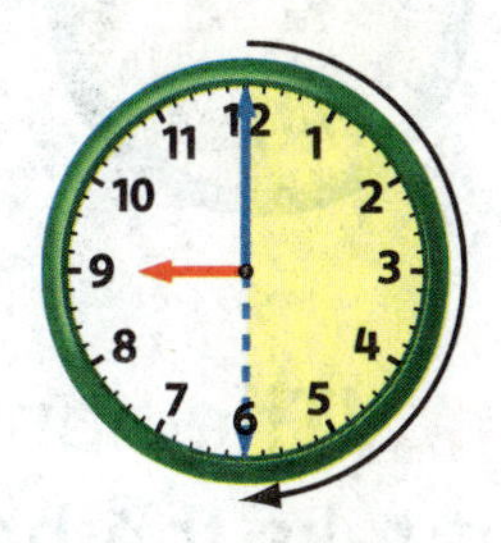

halves Two equal parts of a whole. Each part is a half of the whole.

mitades Dos partes iguales de un todo. Cada parte es la mitad de un todo.

hexagon A 2-dimensional shape that has six sides.

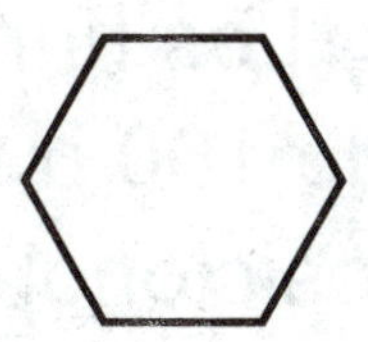

hexágono Figura bidimensional que tiene seis lados.

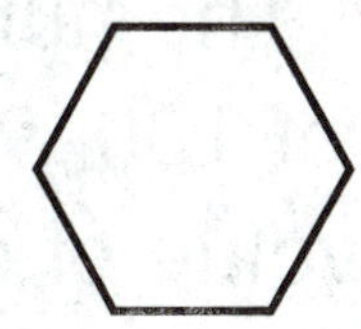

Hh

hour A unit to measure time.

1 hour = 60 minutes

hora Unidad para medir tiempo.

1 hora = 60 minutos

hour hand The hand on a clock that tells the hour. It is the shorter hand.

manecilla horaria Manecilla del reloj que indica la hora. Es la más corta.

hundreds The numbers in the range of 100–999. It is the place value of a number.

365

3 is in the hundreds place.
6 is in the tens place.
5 is in the ones place.

centenas Los números en el rango de 100 a 999. Es el valor posicional de un número.

365

3 está en el lugar de las centenas.
6 está en el lugar de las decenas.
5 está en el lugar de las unidades.

inch (in) A customary unit for measuring length. The plural is *inches*.

12 inches = 1 foot

pulgada (pulg) Unidad usual para medir longitud.

12 pulgadas = 1 pie

inverse Operations that are opposite of each other.

Addition and subtraction are inverse or opposite operations.

operaciones inversas Operaciones que se oponen una a otra.

La suma y la resta son operaciones inversas u opuestas.

key Tells what (or how many) each symbol stands for.

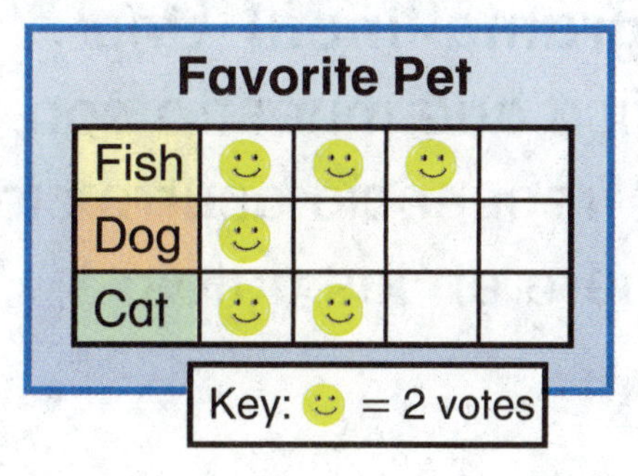

clave Indica qué o cuánto representa cada símbolo.

Animal doméstico favorito

Pez	☺	☺	☺	
Perro	☺			
Gato	☺	☺		

Clave: ☺ = 2 votos

length How long or how far away something is.

length

longitud El largo de algo o lo lejos que está.

longitud

less than <

$4 < 7$

4 is less than 7.

menor que <

$4 < 7$

4 es menor que 7.

line plot A graph that shows how often a certain number occurs in data.

diagrama lineal Una gráfica que muestra con qué frecuencia ocurre cierto número en los datos.

Mm

measure To find the length, height, weight, capacity, or temperature using standard or nonstandard units.

medir Hallar la longitud, altura, peso, capacidad o temperatura usando unidades estándares o no estándares.

meter (m) A metric unit for measuring length.

1 meter = 100 centimeters

metro (m) Unidad métrica para medir longitud.

1 metro = 100 centímetros

minute (min) A unit to measure time. Each tick mark is one minute.

1 minute = 60 seconds

minuto (min) Unidad para medir tiempo. Cada marca es un minuto.

1 minuto = 60 segundos

minute hand The longer hand on a clock that tells the minutes.

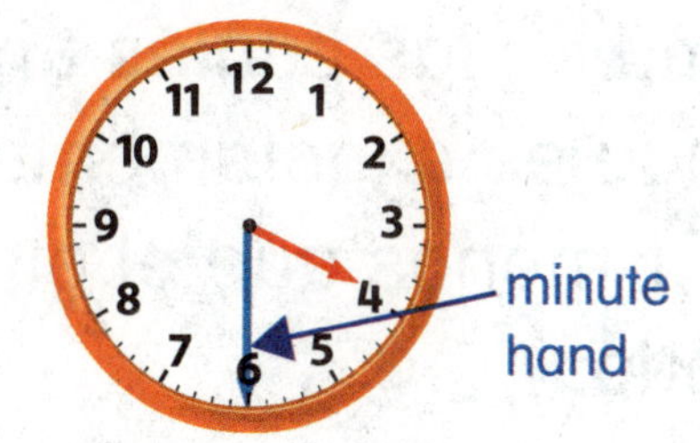

minutero La manecilla más larga del reloj. Indica los minutos.

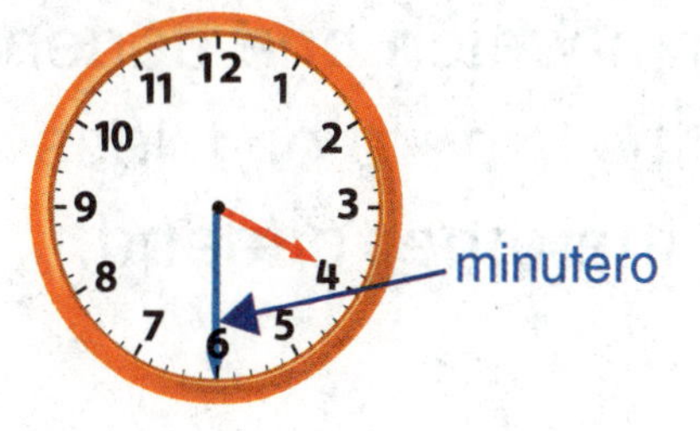

Mm

missing addend The missing number in a number sentence that makes the number sentence true.

$9 + \square = 16$

The missing addend is 7.

sumando desconocido El número desconocido en un enunciado numérico que hace que este sea verdadero.

$9 + \square = 16$

El sumando desconocido es 7.

month A unit of time.
12 months = 1 year

April

Sunday	Monday	Tuesday	Wednesday	Thursday	Friday	Saturday
		1	2	3	4	5
6	7	8	9	10	11	12
13	14	15	16	17	18	19
20	21	22	23	24	25	26
27	28	29	30			

This is the month of April.

mes Unidad de tiempo.
12 mesas = 1 año

Abril

domingo	lunes	martes	miércoles	jueves	viernes	sábado
		1	2	3	4	5
6	7	8	9	10	11	12
13	14	15	16	17	18	19
20	21	22	23	24	25	26
27	28	29	30			

Este es el mes de abril.

Nn

near doubles Addition facts in which one addend is exactly 1 more or 1 less than the other addend.

casi dobles Operaciones de suma en las cuales un sumando es exactamente 1 más o 1 menos que el otro sumando.

nickel nickel = 5¢ or 5 cents

head

tail

moneda de 5¢ moneda de cinco centavos = 5¢ o 5 centavos

cara

cruz

number line A line with number labels.

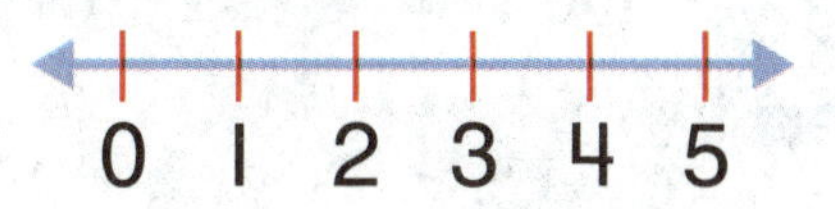

recta numérica Recta con marcas de números.

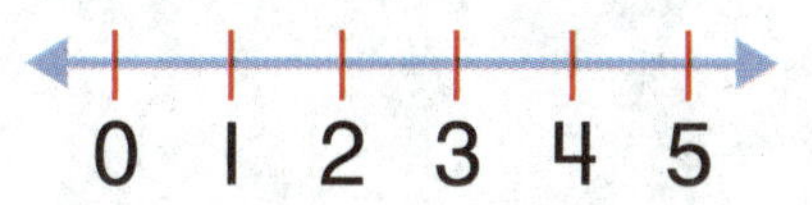

o'clock At the beginning of the hour.

It is 7 o'clock.

en punto El momento en que comienza cada hora.

Son las 7 en punto.

odd number Numbers that end with 1, 3, 5, 7, 9.

números impares Los números que terminan en 1, 3, 5, 7, 9.

Oo

ones The numbers in the range of 0-9. A place value of a number.

65

5 is in the ones place.

unidades Los números en el rango de 0 a 9. Valor posicional de un número.

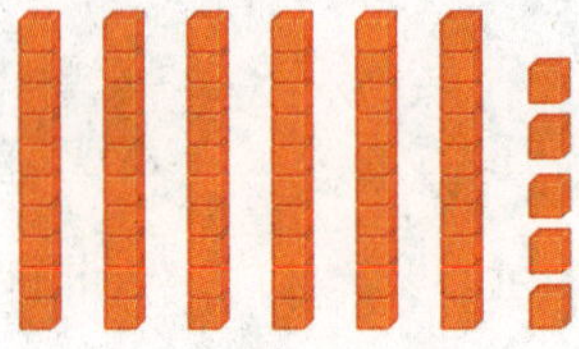

65

El 5 está en el lugar de las unidades.

order

1, 3, 6, 8, 10

These numbers are in order from least to greatest.

orden

1, 3, 6, 8, 10

Estos números están en orden de menor a mayor.

Pp

P.M. The hours from noon until midnight.

p.m. Las horas que van desde el mediodía hasta la medianoche.

parallelogram A two-dimensional shape that has four sides. Each pair of opposite sides is equal and parallel.

paralelogramo Figura bidimensional que tiene cuatro lados. Cada par de lados opuestos son iguales y paralelos.

partition To divide or "break up."

separar Dividir o desunir.

pattern An order that a set of objects or numbers follows over and over.

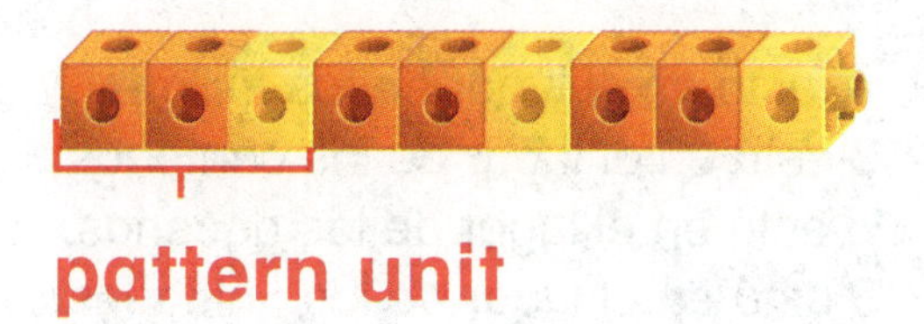

patrón Orden que sigue continuamente un conjunto de objetos o números insertar punto.

penny penny = 1¢ or 1 cent

head

tail

moneda de 1¢ moneda de un centavo = 1¢ o 1 centavo

cara

escudo

pentagon A polygon with five sides.

pentágono Polígono de cinco lados.

picture graph A graph that has different pictures to show information collected.

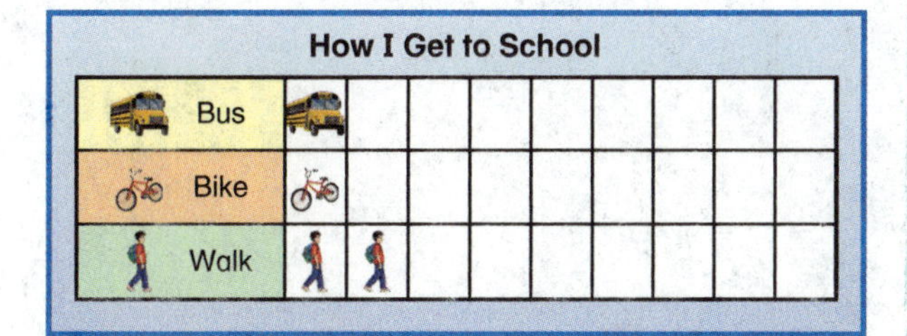

gráfica con imágenes Gráfica que tiene diferentes imágenes para ilustrar la información recopilada.

Cómo voy a la escuela

En Autobús	🚌								
En Bicicleta	🚲								
Caminando	🚶	🚶							

Pp

place value The value given to a *digit* by its place in a number.

1,365

1 is in the thousands place.
3 is in the hundreds place.
6 is in the tens place.
5 is in the ones place.

valor posicional El valor dado a un *dígito* según su posición en un número.

1,365

1 está en el lugar de los millares.
3 está en el lugar de las centenas.
6 está en el lugar de las decenas.
5 está en el lugar de las unidades.

pyramid A three-dimensional shape with a polygon as a base and other faces that are triangles.

pirámide Figura tridimensional con un polígono como base y otras caras que son triángulos.

Qq

quadrilateral A shape that has 4 sides and 4 angles.

cuadrilátero Figura con 4 lados y 4 ángulos.

quarter quarter = 25¢ or 25 cents

tail

moneda de 25¢ moneda de 25 centavos = 25¢ o 25 centavos

cara cruz

quarter hour A quarter hour is 15 minutes. Sometimes called *quarter past* or *quarter til*.

cuarto de hora Un cuarto de hora es 15 minutos. A veces se dice hora y cuarto.

rectangle A plane shape with four sides and four corners.

rectángulo Figura plana con cuatro lados y cuatro esquinas.

rectangular prism A three-dimensional shape with 6 faces that are rectangles.

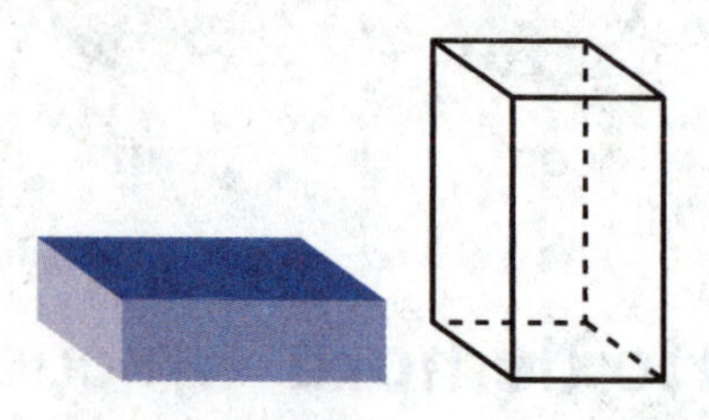

prisma rectangular Figura tridimensional con 6 caras que son rectángulos.

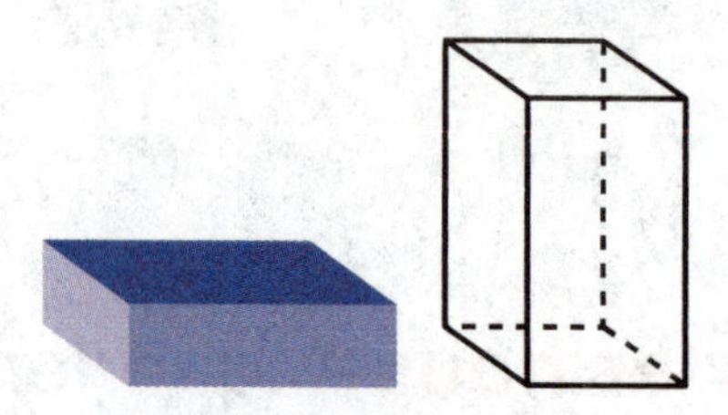

regroup Take apart a number to write it in a new way.

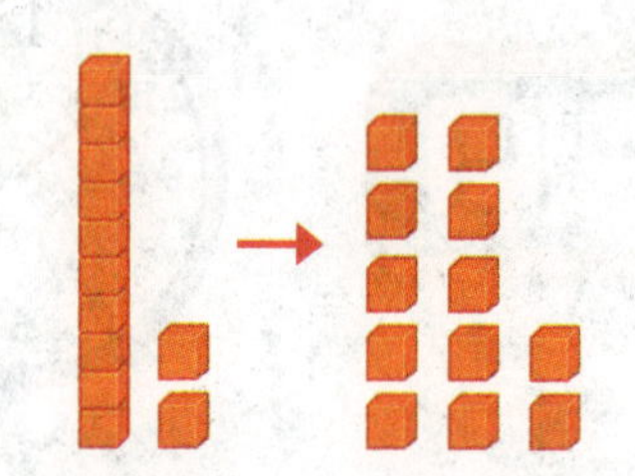

1 ten + 2 ones becomes 12 ones.

reagrupar Separar un número para escribirlo de una nueva forma.

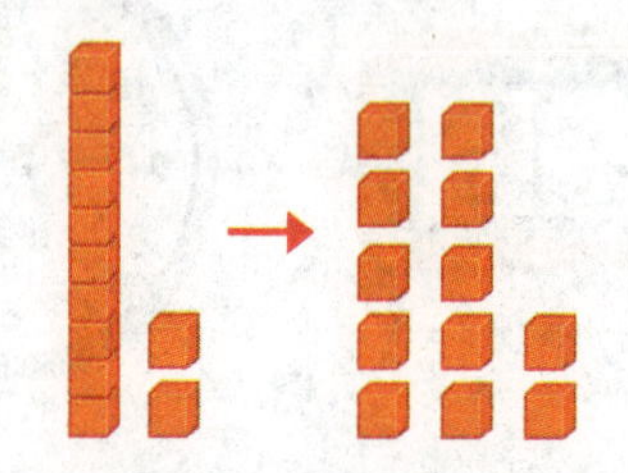

1 decena + 2 unidades se convierten en 12 unidades.

related fact(s) Basic facts using the same numbers. Sometimes called a *fact family*.

$4 + 1 = 5$ $\quad$ $5 - 4 = 1$

$1 + 4 = 5$ $\quad$ $5 - 1 = 4$

operaciones relacionadas Operaciones básicas en las que se usan los mismos números. También se llaman *familias de operaciones*.

$4 + 1 = 5$ $\quad$ $5 - 4 = 1$

$1 + 4 = 5$ $\quad$ $5 - 1 = 4$

repeated addition To use the same addend over and over.

suma repetida Usar el mismo sumando una y otra vez.

rhombus A shape with 4 sides of the same length.

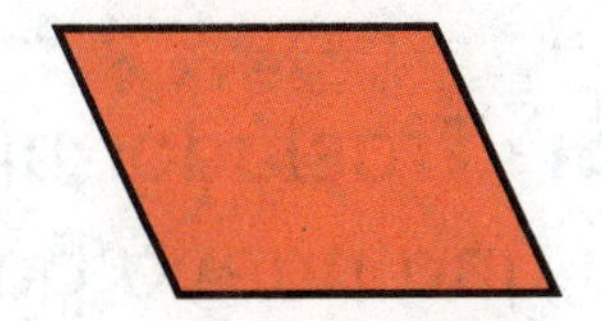

rombo Paralelogramo con cuatro lados de la misma longitud.

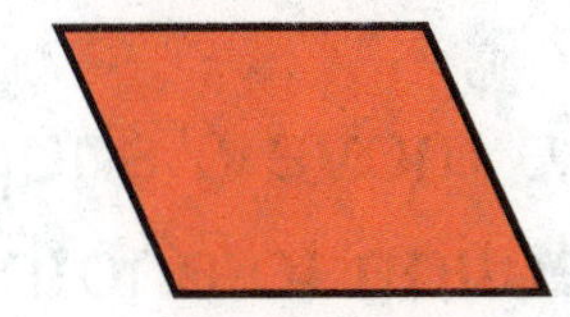

round Change the *value* of a number to one that is easier to work with.

24 rounded to the nearest ten is 20.

redondear Cambiar el *valor* de un número a uno con el que es más fácil trabajar.

24 redondeado a la decena más cercana es 20.

side One of the line segments that make up a shape.

A pentagon has five sides.

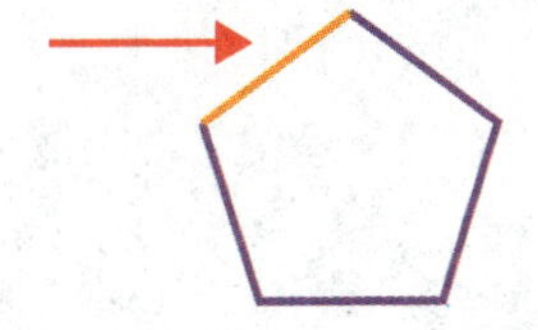

lado Uno de los segmentos de recta que componen una figura.

El pentágono tiene cinco lados.

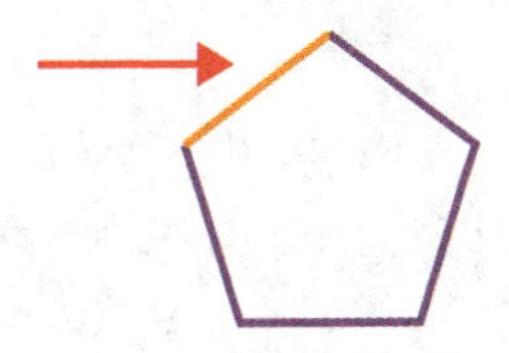

Ss

skip count Count objects in equal groups of two or more.

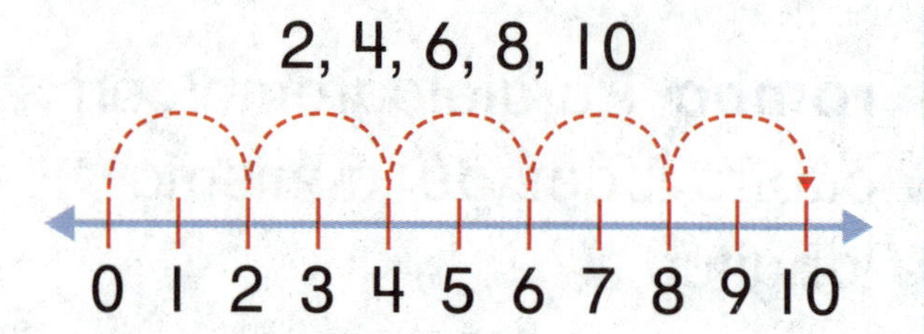

contar salteado Contar objetos en grupos iguales de dos o más.

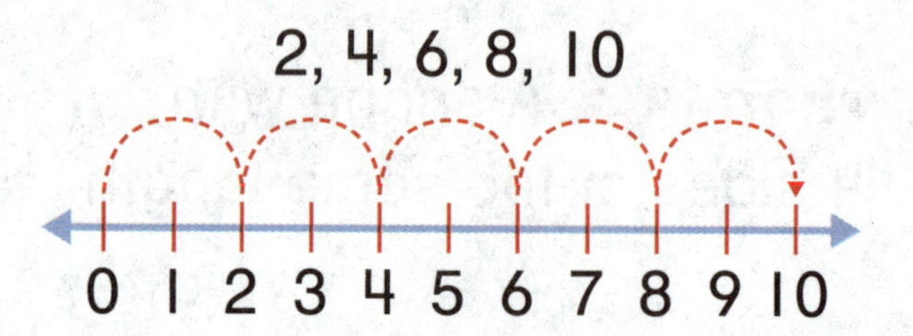

slide To move a shape in any direction to another place.

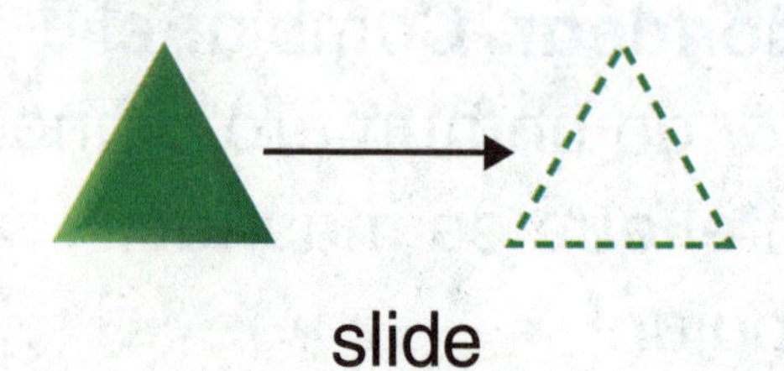

slide

deslizar Traslador una figura a una nueva posición.

deslizar

sphere A three-dimensional shape that has the shape of a round ball.

esfera Figura tridimensional que tiene la forma de una pelota redonda.

square A two-dimensional shape that has four equal sides. Also a rectangle.

cuadrado Figura bidimensional que tiene cuatro lados iguales. También es un rectángulo.

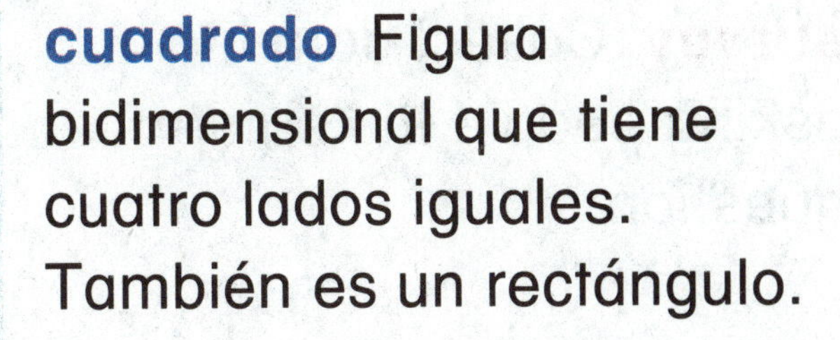

subtract (subtraction) Take away, take apart, separate, or find the difference between two sets. The opposite of *add*.

$5 - 5 = 0$

restar (sustracción) Eliminar, quitar, separar o hallar la diferencia entre dos conjuntos. Lo opuesto de *sumar*.

$5 - 5 = 0$

sum The answer to an addition problem.

$2 + 4 = 6$

suma Resultado de la operación de sumar.

$2 + 4 = 6$

Ss

survey Collect data by asking people the same question.

Favorite Animal	
Dog	𝍸 |
Cat	𝍸

This survey shows favorite animals.

encuesta Recopilar datos al hacer las mismas preguntas a un grupo de personas.

Animal favorito	
Perro	𝍸 |
Gato	𝍸

Esta encuesta muestra los animales favoritos.

symbol A letter or figure that stands for something.

This symbol means to add.

símbolo Letra o figura que representa algo.

Este símbolo significa sumar.

tally marks A mark used to record data collected in a survey.

tally marks

marca de conteo Marca que se utiliza para registrar los datos recopilados en una encuesta.

marcas de conteo

tens The numbers in the range of 10–99. It is the place value of a number.

65

6 is in the tens place.
5 is in the ones place.

decenas Los números en el rango de 10 a 99. Es el valor posicional de un número.

65

6 está en el lugar de las decenas.
5 está en el lugar de las unidades.

thirds Three equal parts.

tercios Tres partes iguales.

thousand(s) The numbers in the range of 1,000–9,999. It is the place value of a number.

1,365

1 is in the thousands place.
3 is in the hundreds place.
6 is in the tens place.
5 is in the ones place.

millar(es) Los números en el rango de 1,000 a 9,999. Es el valor posicional de un número.

1,365

1 está en el lugar de los millares.
3 está en el lugar de las centenas.
6 está en el lugar de las decenas.
5 está en el lugar de las unidades.

three-dimensional shape A shape that has length, width, and height.

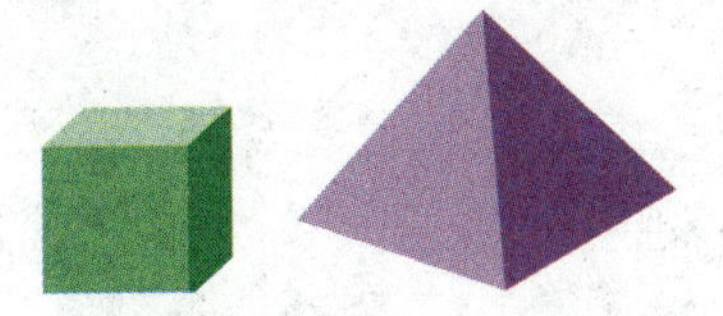

figura tridimensional Que tiene tres dimensiones: largo, ancho y alto.

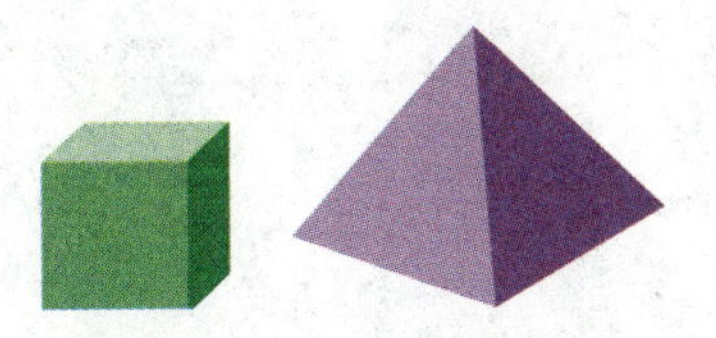

Tt

trapezoid A two-dimensional shape with four sides and only two opposite sides that are parallel.

trapecio Figura bidimensional de cuatro lados con solo dos lados opuestos que son paralelos.

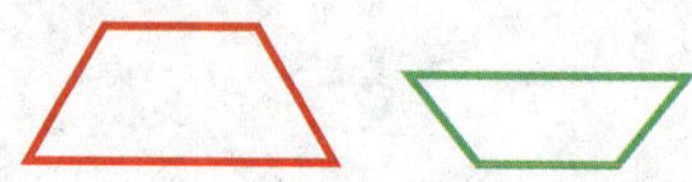

triangle A three-dimensional shape with three sides and three angles.

triángulo Figura tridimensional con tres lados y tres ángulos.

two-dimensional shape The outline of a shape - such as a triangle, square, or rectangle - that has only *length* and *width*.

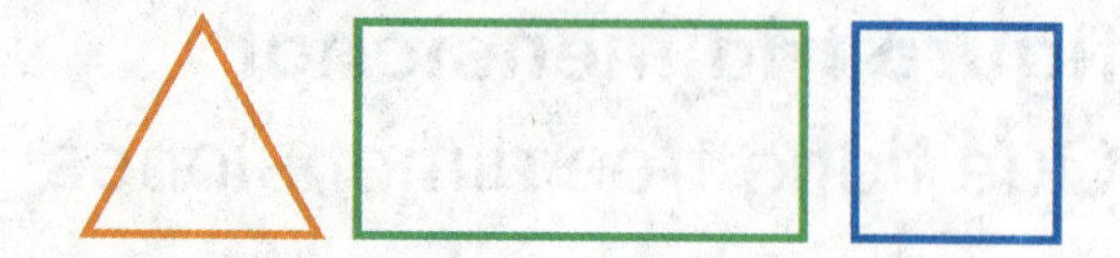

figura bidimensional Contorno de una figura, como un triángulo, un cuadrado o un rectángulo, que solo tiene *largo* y *ancho*.

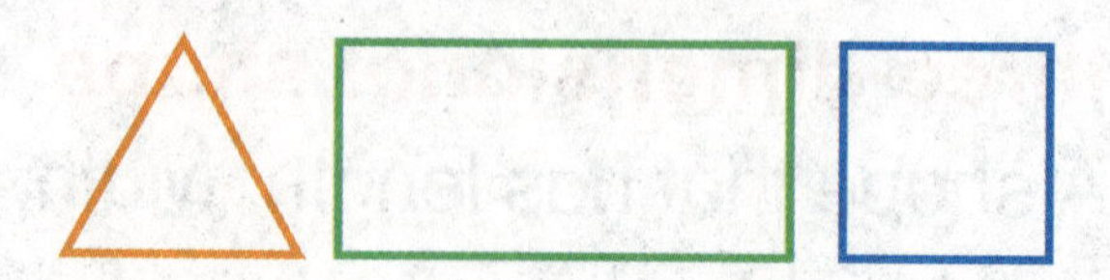

Vv

vertex

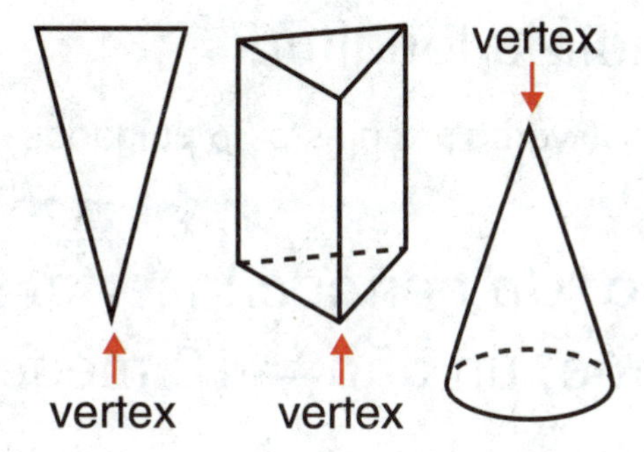

vértice

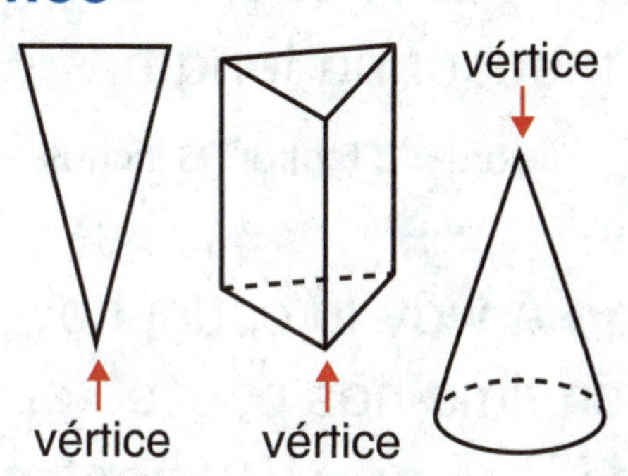

week A part of a calendar.
1 week = 7 days

semana Parte de un calendario
una semana = 7 días

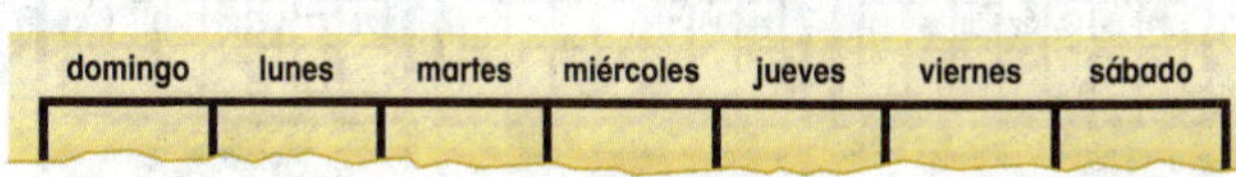

whole The entire amount or object.

el todo La cantidad total o el objeto completo.

yard (yd) A customary unit for measuring length.

1 yard = 3 feet or 36 inches

yarda Unidad usual para medir la longitud.

1 yarda = 3 pies o 36 pulgadas

year A way to count how much time has passed or will pass. 1 year = 12 months

January

S	M	T	W	T	F	S
						1
2	3	4	5	6	7	8
9	10	11	12	13	14	15
16	17	18	19	20	21	22
23	24	25	26	27	28	29
30	31					

February

S	M	T	W	T	F	S
		1	2	3	4	5
6	7	8	9	10	11	12
13	14	15	16	17	18	19
20	21	22	23	24	25	26
27	28					

March

S	M	T	W	T	F	S
		1	2	3	4	5
6	7	8	9	10	11	12
13	14	15	16	17	18	19
20	21	22	23	24	25	26
27	28	29	30	31		

April

S	M	T	W	T	F	S
					1	2
3	4	5	6	7	8	9
10	11	12	13	14	15	16
17	18	19	20	21	22	23
24	25	26	27	28	29	30

May

S	M	T	W	T	F	S
1	2	3	4	5	6	7
8	9	10	11	12	13	14
15	16	17	18	19	20	21
22	23	24	25	26	27	28
29	30	31				

June

S	M	T	W	T	F	S
			1	2	3	4
5	6	7	8	9	10	11
12	13	14	15	16	17	18
19	20	21	22	23	24	25
26	27	28	29	30		

July

S	M	T	W	T	F	S
					1	2
3	4	5	6	7	8	9
10	11	12	13	14	15	16
17	18	19	20	21	22	23
24	25	26	27	28	29	30
31						

August

S	M	T	W	T	F	S
	1	2	3	4	5	6
7	8	9	10	11	12	13
14	15	16	17	18	19	20
21	22	23	24	25	26	27
28	29	30	31			

September

S	M	T	W	T	F	S
				1	2	3
4	5	6	7	8	9	10
11	12	13	14	15	16	17
18	19	20	21	22	23	24
25	26	27	28	29	30	

October

S	M	T	W	T	F	S
						1
2	3	4	5	6	7	8
9	10	11	12	13	14	15
16	17	18	19	20	21	22
23	24	25	26	27	28	29
30	31					

November

S	M	T	W	T	F	S
		1	2	3	4	5
6	7	8	9	10	11	12
13	14	15	16	17	18	19
20	21	22	23	24	25	26
27	28	29	30			

December

S	M	T	W	T	F	S
				1	2	3
4	5	6	7	8	9	10
11	12	13	14	15	16	17
18	19	20	21	22	23	24
25	26	27	28	29	30	31

año Un período insertar punto, un año = 12 meses

enero

D	L	M	M	J	V	S
						1
2	3	4	5	6	7	8
9	10	11	12	13	14	15
16	17	18	19	20	21	22
23	24	25	26	27	28	29
30	31					

febrero

D	L	M	M	J	V	S
		1	2	3	4	5
6	7	8	9	10	11	12
13	14	15	16	17	18	19
20	21	22	23	24	25	26
27	28					

marzo

D	L	M	M	J	V	S
		1	2	3	4	5
6	7	8	9	10	11	12
13	14	15	16	17	18	19
20	21	22	23	24	25	26
27	28	29	30	31		

abril

D	L	M	M	J	V	S
					1	2
3	4	5	6	7	8	9
10	11	12	13	14	15	16
17	18	19	20	21	22	23
24	25	26	27	28	29	30

mayo

D	L	M	M	J	V	S
1	2	3	4	5	6	7
8	9	10	11	12	13	14
15	16	17	18	19	20	21
22	23	24	25	26	27	28
29	30	31				

junio

D	L	M	M	J	V	S
			1	2	3	4
5	6	7	8	9	10	11
12	13	14	15	16	17	18
19	20	21	22	23	24	25
26	27	28	29	30		

julio

D	L	M	M	J	V	S
					1	2
3	4	5	6	7	8	9
10	11	12	13	14	15	16
17	18	19	20	21	22	23
24	25	26	27	28	29	30
31						

agosto

D	L	M	M	J	V	S
	1	2	3	4	5	6
7	8	9	10	11	12	13
14	15	16	17	18	19	20
21	22	23	24	25	26	27
28	29	30	31			

septiembre

D	L	M	M	J	V	S
				1	2	3
4	5	6	7	8	9	10
11	12	13	14	15	16	17
18	19	20	21	22	23	24
25	26	27	28	29	30	

octubre

D	L	M	M	J	V	S
						1
2	3	4	5	6	7	8
9	10	11	12	13	14	15
16	17	18	19	20	21	22
23	24	25	26	27	28	29
30	31					

noviembre

D	L	M	M	J	V	S
		1	2	3	4	5
6	7	8	9	10	11	12
13	14	15	16	17	18	19
20	21	22	23	24	25	26
27	28	29	30			

diciembre

D	L	M	M	J	V	S
				1	2	3
4	5	6	7	8	9	10
11	12	13	14	15	16	17
18	19	20	21	22	23	24
25	26	27	28	29	30	31

Name

Work Mat 1: Ten-Frame

Work Mat 2: Ten-Frames

Name ....

Work Mat 3: Number Lines

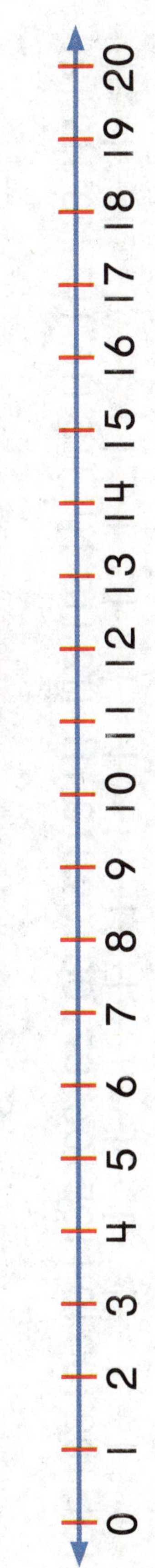

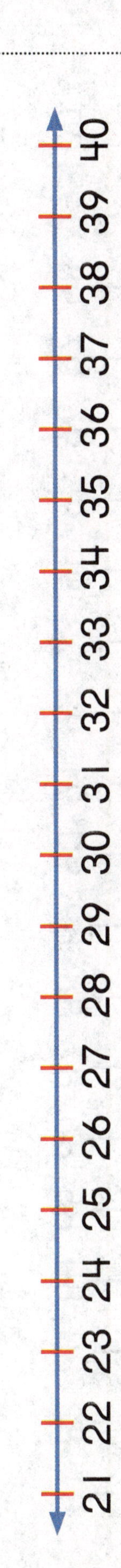

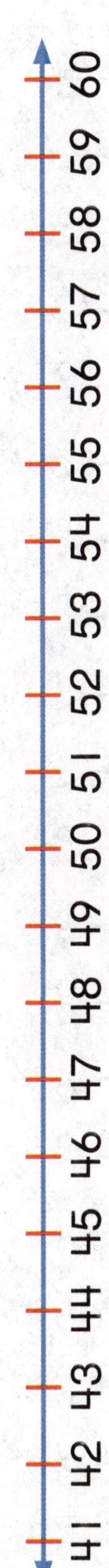

Work Mat 4: Number Lines

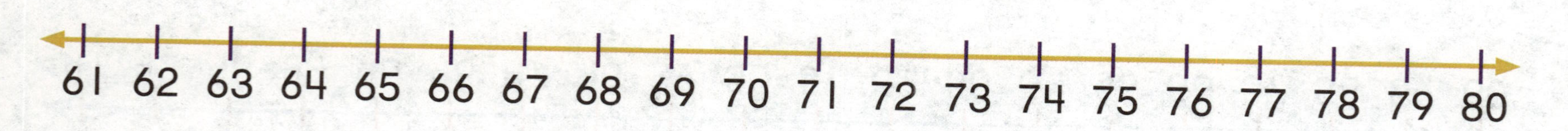

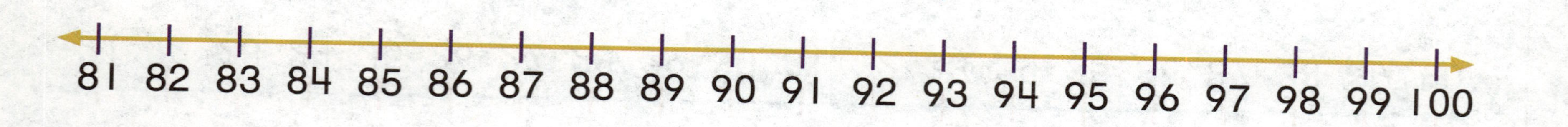

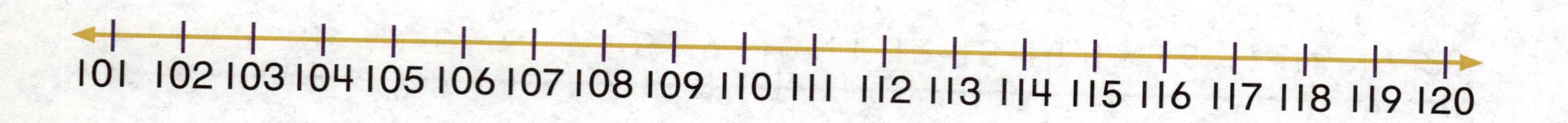

Name

Work Mat 5: Part-Part-Whole

Part	Part

Whole

Work Mat 6: Tens and Ones Chart

Tens	Ones

Name ....

Work Mat 7: Hundreds, Tens, and Ones Chart

Hundreds	Tens	Ones

Work Mat 8: Thousands, Hundreds, Tens, and Ones Chart

Thousands	Hundreds	Tens	Ones